Mathematische Optimierung und Wirtschaftsmathematik | Mathematical Optimization and Economathematics

Reihe herausgegeben von

Ralf Werner, Institut für Mathematik, Universität Augsburg Institut für Mathematik, Augsburg, Bayern, Deutschland

Tobias Harks, Professor für Optimierung, Universität Passau, Passau, Deutschland

Vladimir Shikhman, Technische Universität Chemnitz, Fakultät für Mathematik, Chemnitz, Deutschland

In der Reihe werden Arbeiten zu aktuellen Themen der mathematischen Optimierung und der Wirtschaftsmathematik publiziert. Hierbei werden sowohl Themen aus Grundlagen, Theorie und Anwendung der Wirtschafts-, Finanz- und Versicherungsmathematik als auch der Optimierung und des Operations Research behandelt. Die Publikationen sollen insbesondere neue Impulse für weitergehende Forschungsfragen liefern, oder auch zu offenen Problemstellungen aus der Anwendung Lösungsansätze anbieten. Die Reihe leistet damit einen Beitrag der Bündelung der Forschung der Optimierung und der Wirtschaftsmathematik und der sich ergebenden Perspektiven.

Maximilian Klein

Nested Simulations: Theory and Application

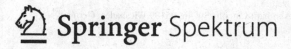
Springer Spektrum

Maximilian Klein
Institute of Mathematics
University of Augsburg
Augsburg, Germany

Zugleich Dissertation, Universität Augsburg, 2023

Datum der Einreichung: 24.02.2023
Tag der mündlichen Prüfung: 15.06.2023
Erster Gutachter: Professor Dr. Ralf Werner
Zweiter Gutachter: Professor Dr. Nikolaus Schweizer
               (Tilburg University)

ISSN 2523-7926   ISSN 2523-7934 (electronic)
Mathematische Optimierung und Wirtschaftsmathematik | Mathematical Optimization
and Economathematics
ISBN 978-3-658-43852-4   ISBN 978-3-658-43853-1 (eBook)
https://doi.org/10.1007/978-3-658-43853-1

This Springer Spektrum imprint is published by the registered company Springer Fachmedien
Wiesbaden GmbH, part of Springer Nature.
The registered company address is: Abraham-Lincoln-Str. 46, 65189 Wiesbaden, Germany

Paper in this product is recyclable.

Für Kathrin,
für all ihre Liebe, Hingabe und Unterstützung.

Danksagung

Eine solche Arbeit entsteht nicht ohne die tatkräftige Unterstützung einiger wichtiger Personen in meinem Leben, ohne die diese Arbeit nicht das geworden wäre, was sie ist.

Zuallererst möchte ich mich recht herzlich bei meinem Doktorvater Prof. Dr. Ralf Werner bedanken. Nicht nur für ein spannendes Thema, sondern vor allem für eine hervorragende Ausbildung und die Möglichkeit sich im Rahmen des Themas frei entfalten zu dürfen. Die Ideen, Anregungen und fachlichen Diskussionen haben mein mathematisches Verständnis nicht nur verbessert, sondern auch entscheidend geprägt.

Des Weiteren möchte ich mich bei Herrn Prof. Dr. Nikolaus Schweizer bedanken, dass er sich dieser Arbeit als Zweitgutachter annimmt.

Ich danke meinen Kolleginnen und Kollegen vom Lehrstuhl für Rechnerorientierte Statistik und Datenanalyse – Kathrin, Christian und Jonas – für die schöne Zeit und für viele anregende fachliche sowie private Diskussionen und Unterhaltungen. Ich habe die Zeit sehr genossen und ohne euch wäre sie nicht ansatzweise so schön gewesen.

Darüber hinaus möchte ich mich bei dem Team von Mayr Investment Managers recht herzlich für die Möglichkeit, neben meiner Promotion wichtige praktische Erfahrungen sammeln zu dürfen, bedanken. Die vielen spannenden und angewandten Projekte sorgten in harten Theoriewochen für eine wichtige Abwechslung. Ein spezieller Dank geht hierbei an Lukas Kretschmer für die tolle Zusammenarbeit und die aufbauenden Worte.

Zu guter Letzt möchte ich mich bei meinen größten Unterstützern bedanken – meinen Eltern, Geschwistern und meiner Frau. Ich danke meinem Vater für das Interesse an meinem Studium und meiner Promotion sowie die Unterstützung über all die Jahre. Ich weiß es sehr zu schätzen, dass er immer an mich und das nächste Ziel geglaubt hat – egal, wie absurd es im ersten Moment klang. Ein weiterer Dank gilt meiner Mutter für ihre nicht endenden Mühen sowie die stets liebevollen und aufmunternden Worte. Von tiefstem Herzen danke ich meinem Stiefvater dafür, den Wunsch nach einer akademischen Ausbildung in mir entfacht zu haben und dass er mir dabei immer mit Rat und Tat zur Seite stand. Abschließend gilt der größte Dank meiner Frau Kathrin, für ihre uneingeschränkte Unterstützung und dass sie während dieser spektakulären Reise immer an meiner Seite stand.

Abstract

In this thesis, we study so called nested Monte Carlo simulations for approximating functions of conditional expected values, i.e. mathematically $X = \mathbb{E}[V|Z]$. Starting with a practical derivation of the problem from actuarial mathematics, the so called Solvency Capital Requirement problem, we transfer it to a general probabilitytheoretical one. Hence, the derived random variables and their behaviour will be analyzed in more detail in the following. In addition, two general problem classes are introduced for the upcoming considerations and analyses: First, the class of momentbased estimators and second, the class of quantile-based estimators. Especially, the consideration of the very general formulated moment-based class leads to an unified and overarching framework compared to mostly explicit considered examples in the literature.

We are the first who analyze the almost sure convergence for moment-based estimators. Thereby, our proofs rely mainly on the complete convergence. The bias of this problem plays a fundamental role, which is why this issue is considered extensively prior to any convergence analyses. Theorem 3.4.1 then summarizes the almost sure convergence of the estimator and its corresponding assumptions. Further, Theorem 3.4.4 provides the underlying rate of convergence of $N^{-1/3}$ in the optimal case, depending on the present bias rate and also the number of existing moments of V.

Afterwards, the almost sure convergence is also proved for the insurance relevant quantile estimators. Similar to the moment-based case, some kind of bias plays a central role for future analyses. Hence, we analyze the deterministic convergence speed between the quantile of the distorted and undistorted X. Based on this, we obtain in Theorem 4.3.1 the first and so far only almost sure convergence result for quantile estimators based on a nested Monte Carlo simulation. This result is then further extended in Theorem 4.3.5 by considering also the underlying convergence rate. Here, we prove an optimal rate of $N^{-1/4}$ under weak assumptions on the moments of V. Compared to the optimal Root Mean Squared Error (RMSE) rate of $N^{-1/3}$, shown by Gordy and Juneja (2010), under much more restrictive smoothness assumptions, this is a remarkable and, to the best of our knowledge, unique result to date.

Furthermore, in addition to the results mentioned so far, we present a novel non parametric confidence interval method for quantiles. Similar to the approach of Lan et al. (2007b), this method takes the noise within the simulation into account, which arises due to the fact of two concatenated simulations. We show that our approach leads on the one hand to a simpler parameterization and on the other hand—according to numerical analyses—to narrower confidence intervals. Furthermore, with this approach it is also possible to obtain additionally an asymptotic error estimate via the Berry-Esseen Theorem. The asymptotic confidence interval method and the corresponding error estimate are summarized in Theorem 5.3.3.

We complement this work with an extensive numerical analysis on the proven theoretical rates and the application of the novel confidence interval method. Here, for the complexity analysis selected academic examples will be considered, since here the respective analytical solution is needed. For the confidence interval analysis the Asset Liability Model (ALM) of Hieber et al. (2019) was used beside an academic test case in order to demonstrate the practicability of the method. All implementations were carried out in MATLAB.

Zusammenfassung

In dieser Arbeit untersuchen wir sogenannte Nested Monte Carlo Simulationen zur Approximation von Funktionen bedingter Erwartungswerte, d. h. mathematisch $X = \mathbb{E}[V|Z]$. Beginnend mit einer praktischen Herleitung der Problemstellung aus der Versicherungsmathematik, dem sogenannten Solvency Capital Requirement Problem, wird das Thema von diesem in ein allgemeines wahrscheinlichkeitstheoretisches überführt, wobei im Folgenden die hierfür abgeleiteten Zufallsvariablen und deren Struktur näher analysiert werden. Darüber hinaus werden für folgende Betrachtungen und Analysen zwei allgemeine Problemklassen eingeführt: Zum einen die Klasse der momentbasierten Schätzer und zum anderen die Klasse der quantilbasierten Schätzer. Vor allem die Betrachtung der allgemein formulierten momentbasierten Klasse führt im Vergleich zu meist explizit betrachteten Beispielen in der Literatur zu einem einheitlichen und übergreifenden Rahmen.

Als Erste in diesem Themenbereich analysieren wir die fast sichere Konvergenz für momentbasierte Schätzer und weisen diese mithilfe der vollständigen Konvergenz nach. Dabei spielt der Bias dieser Problemstellung eine fundamentale Rolle, weswegen diese Fragestellung ausgiebig vor etwaigen Konvergenzanalysen betrachtet wird. Theorem 3.4.1 fasst dann die fast sichere Konvergenz des Schätzers und deren Annahmen zusammen. In Theorem 3.4.4 wird darüber hinaus noch die zugrundeliegende Konvergenzgeschwindigkeit nachgewiesen, welche in Abhängigkeit der vorliegenden Biasgeschwindigkeit und der Anzahl an existierenden Momenten von V im optimalen Fall zu einer Rate von $N^{-1/3}$ führt.

Anschließend wird die fast sichere Konvergenz ebenfalls für die versicherungsmathematisch relevanten Quantilsschätzer nachgewiesen. Ähnlich wie im momentbasierten Fall spielt für künftige Analysen auch hier eine Art Bias eine zentrale Rolle. Dabei wird die deterministische Konvergenzgeschwindigkeit zwischen dem verzerrten und unverzerrten Quantil der Zielgröße X analysiert. Basierend auf dieser erhalten wir in Theorem 4.3.1 als Erste und bisher Einzige eine fast sichere Konvergenz für Quantilsschätzer basierend auf einer Nested Monte Carlo Simulation. Dieses Resultat wird dann in Theorem 4.3.5 noch durch die zugrundeliegende Konvergenzgeschwindigkeit erweitert. Hier weisen wir eine optimale Rate von $N^{-1/4}$ unter schwachen Annahmen an die Momente von V nach. Verglichen mit der von Gordy and Juneja (2010) nachgewiesenen optimalen Root Mean Squared Error (RMSE) Rate von $N^{-1/3}$ unter deutlich restriktiveren Stetigkeitsannahmen ist dies ein bemerkenswertes und bis jetzt nach unserem besten Wissen einzigartiges Resultat.

Des Weiteren stellen wir neben den bisher genannten Resultaten noch eine neuartige, nicht parametrische Konfidenzintervallmethode für Quantile vor. Diese berücksichtigt, ähnlich wie der Ansatz von Lan et al. (2007b), das Rauschen innerhalb der Simulation, welches aufgrund von zwei verschachtelten Simulationen entsteht. Dabei sei hervorzuheben, dass unser Ansatz zum einen zu einer einfacheren Parametrisierung und zum anderen – anhand von numerischen Analysen – zu engeren Konfidenzintervallen führt. Darüber hinaus ist es mit diesem Ansatz möglich eine asymptotische Fehlerabschätzung über den Satz von Berry-Esseen abzuleiten. Die asymptotische Konfidenzintervallmethode und die dazugehörige Fehlerabschätzung sind in Theorem 5.3.3 zusammengefasst.

Wir runden bzw. schließen diese Arbeit mit einer ausgiebigen numerischen Analyse über die nachgewiesenen theoretischen Raten und der Anwendung der neuartigen Konfidenzintervallmethode ab. Hierbei werden für die Komplexitätsanalyse ausgewählte akademische Beispiele betrachtet, da hier die jeweilige analytische Lösung von Nöten ist. Für die Analyse der Konfidenzintervalle wurde auf das Asset Liability Model (ALM) von Hieber et al. (2019) zurückgegriffen und die Praktikabilität der Methode nachgewiesen. Alle Implementierungen wurden in MATLAB durchgeführt.

Contents

List of Figures

List of Tables

Introduction

1.1   Motivation

As the English economist John Kay described in an interview with Zeit[1] as well
as in his book Kay and King (2020), a fundamental examination of the concept of
risk in the modern financial and insurance world is indispensable. Unfortunately,
the concept of 'radical uncertainty', grounding on a non-stationary environment in
economics and social sciences, must be admitted. The economic and financial crisis
in 2008 and 2010 (Greek government-debt crisis) made unmistakably clear that a
profound understanding of risk and thus models for measuring risk are of extreme
importance. However, the belief that such models describe 'the world as it really is'
is deceptive, but they can give helpful advice to avoid past errors and are an useful
compass for unknown developments under 'radical uncertainty'.

In this respect, risk assessment and management plays a crucial role in business
and also our daily life. Aven (2016) describes it as rapidly changing and still 'shaky'
evolving field of research over the past few decades with the goal to understand the
'world' not exactly as it really is but in such a way that it is 'manageable'. In the
financial and insurance sector a key development of risk assessment lies in the
established stochastic risk models. At this point comes the manageable part into
play because future is, generally, from today onwards uncertain or maybe somehow
'radical uncertain' and can thus, by non-stochastic models, only be described based
on past observations. Now, the crucial part is, first and foremost, the identification
of the most relevant risk drivers of the underlying business which describes the
risk under consideration, cf. Aven (2016). Then, a stochastic risk model generates

---

[1] T. Fischermann. Chance oder Risiko? Die Zeit, Vol. 31, 2019.

© The Author(s), under exclusive license to Springer Fachmedien Wiesbaden GmbH,
part of Springer Nature 2024
M. Klein, *Nested Simulations: Theory and Application*, Mathematische Optimierung
und Wirtschaftsmathematik | Mathematical Optimization and Economathematics,
https://doi.org/10.1007/978-3-658-43853-1_1

a large scenario set of potential realizations of such risk drivers. These try, based on the Glivenko-Cantelli Theorem, to cover possible future outcomes randomly. Second, the evaluation and the associated question of how risk can be quantified has to be answered. Therefore, risk measures gained attention during the last decades for assigning a real number to a risk. In this thesis, so called moment- and quantile-based risk measures are of particular interest.

Consequently, to such stochastic quantile-based risk approaches, the European Parliament and European Council (2009) passed the Solvency II directive in 2009, whereby the final implementation in the individual states of the EU still dragged until 2016. One main objective was that each European insurer is obliged to reserve enough equity to protect, first of all, each policyholder and second the general public from insolvencies of insurance institutions. If the aftermath of the financial crisis figured out one thing clearly, then that insolvency of major financial institutions can lead to severe capital market dislocations and thus to expensive government bailouts. This shocking scenario which occurred in the banking sector (due to the bankruptcy of Lehman Brothers) should by no means be repeated for insurance institutions. As general application this dissertation focuses on the life insurance sector and hence by insurers we directly imply life insurance companies. The Solvency II directive introduces a new quantile-based risk buffer, named Solvency Capital Requirement (SCR), which aims on the described objective of higher equity numbers. In this regard, insurers are obliged to derive the market values of their underlying balance sheet for a one-year time horizon. Thus, market-consistent valuation approaches for life insurance contracts and therefore adequate tools for pricing assets and liabilities are needed.

1.2   Context

As indicated in Section 1.1, the main lesson learned from the financial crisis and earlier years was the need for a regulation that increases particularly the reserves of financial institutions based on stochastic risk models. For banks, this was achieved via Basel II and for insurance companies via Solvency II.

Economic Balance Sheet, Available Capital & SCR

The Solvency II guideline defines the SCR as the 0.5%-quantile, resp. Value-at-Risk (VaR), of the Available Capital (AC) or equivalently, to be in line with the Solvency II terminology, of the Basic own Funds (BoF) over a one year period. The BoF in this context refers to the amount which can be used as risk buffer having deduced the underlying obligations against policyholders and stakeholders. Hence, in our academic setting with an assumed simplified balance sheet (cf. Figure 1.1) the BoF is given as the difference between the market value of assets and liabilities. Hence, it holds

$$\text{MV}^A = \text{MV}^L + \text{BoF}. \tag{1.1}$$

The risk management process can, then, be divided into two separate components. First, deriving the BoF at present (i.e. $t = t_0$) and second the SCR grounding on the BoF for the upcoming period t_1 (according to Solvency II for one year). Note that the liability side of a life insurance company provides some unique specialties like e.g. the long duration of the policies or possible mortality. This work focuses on the market value of liabilities and its derivation approaches because the underlying assets and its market value derivation are very common in financial mathematical terms and mostly analytical formulas or adequate numerical solvers are available.

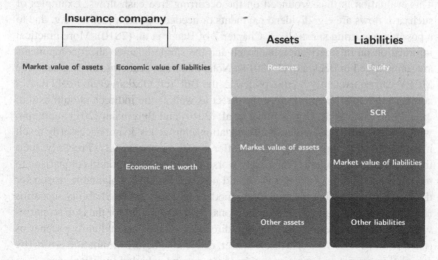

Figure 1.1 Simplified balance sheet of an European insurer (cf. Hancock et al. (2001))

Moreover, the BoF or more precisely the liability cash flows can be derived either from a direct or an indirect perspective. In our terminology the indirect method, on the one hand, takes a policyholder viewpoint and evaluates the occurring liability cash flows. Hence, the insurance business and in particular the underlying life insurance policies must be simulated and will then be used for BoF evaluations. This modeling approach will be discussed in Section 1.3. On the other hand, the calculation can also be seen from the opposite angle and thus from a shareholder perspective since tax particularities are not considered here. This approach is called the direct method and aims in evaluating the corresponding BoF values instead of the liabilities. The derivation is grounded on the principles of the European Insurance CFO Forum, in short CFO Forum, a consortium consisting of the Chief-Financial-Officers (CFOs) of major European insurance companies. Its aim is to influence regulatory developments for European insurers and implementing them—based on current research developments—in a way that is coherently manageable for all institutions. From a shareholder, analyst, investor, regulatory or management perspective a significant key value of an insurance company is a general and especially insurance wide comparable measure of profitability. Hence, the CFO Forum introduced the Market Consistent Embedded Value (MCEV) and the Present Value of Future Profits (PVFP) for an European wide comparison, see CFO Forum (2009, 2016). This evaluation is thus grounded on the occurring free cash flows. Examples of such cash flows are e.g. dividend payments or needed capital injections e.g. due to a possible occurring shortfall, cf. Chapter 7 of Bauer et al. (2010). More practical information on this derivation approach and the single balance sheet components are summarized in Becker et al. (2014). Note, that in an academic environment the MCEV can equivalently be interpreted as the BoF (cf. Oezkan et al. (2011)). According to (1.1) both approaches, the direct as well as the indirect, should lead to exactly the same SCR. But as Bauer et al. (2010) and Bergmann (2011) underpin with several numerical studies, both evaluation approaches do not necessarily result in a similar SCR approximation and differ sometimes extremely. Precisely, their indirect method leads to a significantly worse SCR approximation in comparison to the direct method. In general, this should not happen. One explainable reason for this error, assuming that the underlying model is able to simulate liability scenarios (indirect method), could be that an additional cash flow model for the BoF scenarios was used, instead of calculating the BoF directly from the present liability scenarios (i.e. $BoF = MV^A - MV^L$). This, however, can lead to severe simulation errors (by cumulation errors) and explains maybe the occurring calculation differences.

Since the MCEV and thus the direct method was in first line introduced as an inter insurance comparison value and not for risk management tasks we focus in the remaining work on the indirect approach which seems also more justified by Solvency II. To derive the underlying key values, we first present the basic model under the assumption that a closed form solution is possible. Later on, we motivate the Monte Carlo approximation within this setup. The BoF at t_0 is in this setting essentially defined as the difference of the market value of assets and liabilities, i.e.

$$\text{BoF}_{t_0} := \text{MV}_{t_0}^A - \text{MV}_{t_0}^L. \tag{1.2}$$

Recall, that this is a completely deterministic quantity because all current asset and liability values can be determined based on present information. Due to the fact of the unknown financial market evolution, the Basic own Funds at t_1 are, in contrast, a stochastic quantity since the underlying risk factors of the insurance policies between t_0 and t_1 depend highly on the upcoming financial situation. Hence, at $t = t_1$, we have

$$\text{BoF}_{t_1} := \text{MV}_{t_1}^A - \text{MV}_{t_1}^L. \tag{1.3}$$

An even more detailed definition with the corresponding cash flows is summarized in Section 1.3. Before defining the SCR we introduce in a next step the underlying loss distribution of an exemplary insurance company. With the BoF definitions the loss between t_0 and t_1 can be illustrated as the difference of BoF_{t_0} and BoF_{t_1}, i.e.

$$\Delta := - \left(\text{BoF}_{t_1} - \text{BoF}_{t_0} \right). \tag{1.4}$$

Note that in practice, the distribution of Δ is commonly unknown and hence we will approximate it later on by an empirical distribution function based on an underlying nested simulation framework. But in this analytical setup the Solvency II guideline defines the SCR as the α-quantile or Value-at-Risk (VaR^L) of the underlying loss distribution for $\alpha = 99.5\%$, i.e.

$$\text{SCR} := \inf \{ c \in \mathbb{R} \mid \mathbb{P} (\Delta > c) \leq 1 - \alpha \} = q_{\alpha,-}^{\Delta} = -q_{1-\alpha,+}^{-\Delta} \tag{1.5}$$
$$= VaR_{\alpha}^L(\Delta) = VaR_{\alpha}^L(-\text{BoF}_{t_1}) + \text{BoF}_{t_0}.$$

Recall, BoF_{t_0} is a deterministic value. Hence, the last equality follows immediately from the cash invariance property of the Value-at-Risk. Here, $q_{\alpha,+}^{\Delta}$ denotes the upper and $q_{\alpha,-}^{\Delta}$ the respective lower α-quantile of the underlying loss distribution Δ, cf. Föllmer and Schied (2016) (Definition A.18). The VaR, however, is defined as the negative upper quantile function of a profit, cf. McNeil et al. (2015) (Definition 2.8) or Föllmer and Schied (2016) (Definition 4.45). Each insurer in the European Union is obliged to withhold this risk buffer for one year in his balance sheet for extreme but possible emerging risk scenarios to avoid an insolvency in the upcoming year based on present business. Thereby, the security level α is predetermined and fixed at 99.5%. Hence, for this quantile level, the definition now states that, according to the underlying risk scenarios, the probability of a next year's loss exceeding the SCR is less than or at most equal to 0.5%. Since BoF_{t_0} has no major conceptual influence on the risk calculation and $-\mathrm{BoF}_{t_1}$ defines a loss, the focus lies in the following on

$$VaR_\alpha^L(-\mathrm{BoF}_{t_1}) = VaR_{1-\alpha}(\mathrm{BoF}_{t_1}) = -q_{1-\alpha,+}^{\mathrm{BoF}_{t_1}},$$

for $\alpha = 99.5\%$. In an actuarial context it is common sense that the underlying loss cdf F_Δ is continuous as well as strictly increasing and thus invertible. Hence, the given quantile definitions coincide, i.e. for all $\alpha \in \mathbb{R}$

$$q_{\alpha,-} = q_\alpha = q_{\alpha,+}$$

holds, see McNeil et al. (2015) (Proposition A.3).

However, after introducing the essential key variables of the risk management process, it is, until now, completely unclear how insurance liability cash flows are modeled. This is one crucial task in risk departments which enables the next step in deriving (1.2) and especially (1.3).

Asset-Liability Models (ALMs)

Asset-Liability Models (ALMs) gained attention in the actuarial community around time when Solvency II was published and proposed the market-consistent valuation approach. The asset and liability cash flows must be, first of all, modeled and, second, in a later step valued. Especially, the liability modeling turns out to be a challenging and also computationally intensive part because life insurance contracts include some peculiarities in contrast to classical financial mathematical setups. Examples are given by the occurring mortality, possible surrender options or withdrawal guarantees. These are just a few special characteristics of life insurance contracts. Hence, the liability cash flows cannot be represented by basic or well-known financial instruments. Obviously, the additional occurring long duration time (terms of more than 40 years are common) results in severe path dependencies and thus leads to extremely complex cash flow structures on the liability side. Hence, analytical evaluations are impossible and skilful simulation approaches are indispensable. Note that the next paragraph introduces some common ALMs and does not intend to present all the modeling details.

Known academic ALMs which address a wide range of practical problems in life insurance portfolios focusing on cash flow simulation are given by Gerstner (2008), Holtz (2008) and Bauer et al. (2012). The model from Bauer et al. (2012) proposes two simulation approaches based on the German life insurance market, a so-called MUST- and IS-case. The MUST-case covers the minimal requirements on ALMs which German life insurers are faced due to regulatory authorities and applicable law, whereas the IS-case describes the common and applied practice in German life insurance companies until 2010. Of course, unique in-house management rules which vary for each insurer, grounding on e.g. the underlying business profile or special and most recent regulatory rules like the 'Zinszusatzreserve' in Germany or other actuarial reserves, are not covered in such a general academic model. Contrary, Gerstner (2008) and the associated dissertation of Holtz (2008) try to capture several of these specialties in an even more detailed simulation model. For instance, various management (e.g. parameters for an investment rate in stocks, a current reserve rate, etc.) and furthermore actuarial reserve parameters are considered.

Summarized, ALMs simulate cash flows of assets and liabilities based on a current state of the financial market in order to mirror the balance sheet of an insurance company. First and foremost, ALMs are primarily suitable for the determination of the BoF for a fixed time point and are hence to short-sighted for the whole Solvency II risk management process. To circumvent this, the underlying cash flows must be projected into the future, precisely reevaluated under various economic scenarios resp. financial market states. Then, it is possible to derive the underlying cash flows of BoF_{t_1} with an ALM. This projection problem will be solved by so called cash flow projection models in the next section.

Cash Flow Projection (CFP)

It should be emphasized that Solvency II provides a standard formula for calculating the risk capital for the upcoming year. Nevertheless, it gives life insurance companies also the freedom to develop their own internal model. By using such an internal model, insurance companies aim to reduce their risk capital, as specific business strategies and internal diversification effects can be better taken into account by an internal model than by the predefined standard formula. Thus, insurers perform an internal inventory of their underlying and particularly future risk drivers at regular intervals—generally at annual intervals—and hence the derivation of future equity numbers e.g. the BoF at t_1 and the SCR derivation. The decisive fact in Solvency II that insurers have to derive their full loss distribution, requires the usage of numerical simulations in risk management departments because the pricing of special features, like surrender options, withdrawal payments etc., in the liability cash flows lead, as mentioned previously, to rather complex payoff structures which cannot be handled by analytical formulas. In practice, such models are summarized under the heading of CFP models. In Section 1.3 we first investigate the analytical setup and then introduce the practically relevant approximation framework over nested Monte Carlo simulations (cf. Section 1.4). For an overview, we present and explain below the procedure in a rather general way to illustrate important key findings of CFP models in the insurance business.

Specified by its definition and in short CFP models are reevaluating/projecting the economic balance sheet of an insurer under various risk factors one year into the future and generate the desired cash flows up to maturity. This is a rather complex and challenging endeavor according to the liability specialties and the longevity. Therefore, a major task of every risk management department is an analysis which risk factors resp. macroeconomic variables drive their current insurance portfolio. Afterwards, first and foremost an economic scenario generator (ESG) will be used to generate and model these specified risks from t_0 to t_1 (we refer to this as the risk horizon). Second, the ESG also models the needed risk factor developments throughout the whole projection horizon (i.e. from $t = t_1$ to maturity date $t = T$). In this regard, the Investment Committee of DAV (2015) discusses various stochastic models – such as the LIBOR market model, the Vasicek model or geometric Brownian motion models—and highlight additionally the calibration as crucial modeling part in ESG simulations. For the sake of simplicity we generally distinguish between two types of risk factors: The evolving risks of the financial market and the underlying actuarial risks. Common financial market risks—faced by probably every life insurer—are e.g. interest rate or stock price changes which directly affect the value of policyholder accounts (by e.g. participating insurance contracts, cf. Hieber et al. (2019)). Actuarial risks, like lapse behaviour or mortality and many more, on the other hand, are particularly driven by historical data or predefined by various industry-wide assumptions, such as mortality tables. Nevertheless, we will not go into modeling details and rely on the simplified ALM framework of Hieber et al. (2019) in the forthcoming numerical section (cf. Chapter 6) of this thesis. Supplementary information on this issue can be found for example in Krah et al. (2018) and the associated dissertation of Krah (2020). Here, the authors explain major practical problems and indicate some realistic risk factors for life insurance portfolios.

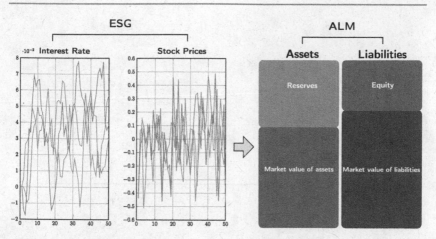

Figure 1.2 Relationship between the stochastic economic scenario generator (ESG) and the asset liability model (ALM)

Based on a generated ESG file an ALM generates/calculates the corresponding asset and liability cash flows for every time point between t_1 and T. Hence, the risk factor projection between t_0 and t_1 takes place according to the generated ESG or risk scenarios. Obviously, the ESG scenarios serve—in a CFP model environment—as input for the ALM (cf. Figure 1.2). More specifically, all asset and especially liability cash flows have to be evaluated conditioned on each generated risk scenario, i.e. under the underlying or modeled financial situation, c.f. Figure 1.3. Thus, the previously mentioned ALM simulates—based on every evolved risk scenario—the asset and liability cash flows up to maturity date T. Furthermore, it takes all possible asset-liability interactions or additional company-specific management rules into account. These cash flows will then be used to derive the fair value of the underlying assets and especially of the liabilities, see in the following Section 1.3.

Having introduced the general SCR values and the corresponding cash flow and projection tools, we have in the remainder of this introductory chapter to clarify the still missing market value evaluation problem in the introduced BoF formula, cf. (1.2) and (1.3). Hence, Section 1.3, first of all, introduces the closed form problem and the necessary framework. Second of all, we then derive an approximation procedure based on Monte Carlo simulations to determine the key values of the risk management process, namely the market values and their associated loss distribution.

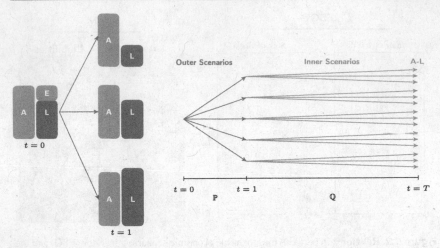

Figure 1.3 Projection of the balance sheet and an uniform nested simulation tree

1.3   Analytical Problem Statement—The Actuarial Perspective

A precise mathematical quantification of the SCR problem and especially the missing market value determination is the intended objective of this section. To this end, we essentially adopt the previous insurance mathematical and simulation framework of Bauer et al. (2010), Bergmann (2011), Natolski (2018) and Krah (2020). Where necessary, these approaches will be extended to the current state of practical implementation or adjusted to our own notation.

Probability Space

In our modeling setup we assume that the cash flows, evolving from liabilities and assets, occur only at discrete points in time, i.e. our time horizon is given by $\mathcal{T} = \{t_0\} \cup \mathcal{T}_p$ with $\mathcal{T}_p = \{t_1, \ldots, T\}$ for $t_0, t_1, \ldots, T \in \mathbb{R}$ in ascending order. Note that $\mathcal{T}_p$ denotes the addressed projection horizon and $\mathcal{T}_r = \{t_0, t_1\}$ the risk horizon. Let $(\Omega, \mathcal{F}, \mathbb{F}, \mathbb{P})$ be a complete filtered probability space, with filtration $\mathbb{F} = (\mathcal{F}_t)_{t \in \mathcal{T}}$ and a real world probability measure $\mathbb{P}$. Each sigma algebra $\mathcal{F}_t \subseteq \mathcal{F}$ covers all knowable information up to the respective time t.

Risk Factors

Grounding on the fact that the future profits of a life insurer are driven by various risk factors, we assume a d-dimensional Markov process $\mathbf{Z} = (\mathbf{Z}_t)_{t \in \mathcal{T}} = (Z_{1,t}, \ldots, Z_{d,t})_{t \in \mathcal{T}} \in \mathbb{R}^d$ and that all underlying assets can be expressed in terms of this process. $\mathbf{Z}$ is an observable stochastic process and generates the filtration $\mathbb{F}$. This process is—in financial mathematics—commonly known as state process. It symbolically describes the current or future state of the risk factors of an insurance company and thus, in particular, the value of the actuarial and financial indicators. They are part of the previously mentioned ESG file. The dimension of the process represents different risk factors. To illustrate this with a few figures Krah et al. (2018), for example, use a model based on 17 risk factors, i.e. $d = 17$. Thereby, nine factors describe the capital market risks and the remaining eight correspond to actuarial risks. Hieber et al. (2019) instead consider a simplified model with $d = 2$ capital risk factors (i.e. stocks and interest rates).

Numéraire

Additional to the state process, we suppose the existence of a positive, adapted stochastic process $\mathbf{N} := (N_t)_{t \in \mathcal{T}}$ with $N_{t_0} = 1$. We also assume the existence of a risk-neutral probability measure $\mathbb{Q}$, equivalent to the real world measure $\mathbb{P}$ with respect to this numéraire $\mathbf{N}$.

Terminal Values (TV) & Discounted Terminal Values (DTV)

As the market value of assets resp. liabilities has to be calculated, an intermediate step is to define, first of all, their corresponding cumulative Terminal Values (TV) and Discounted Terminal Values (DTV) based on the underlying cash flow projections arising from the available CFP model. Therefore, we introduce two separate adapted cash flow processes $\left(CF_t^A\right)_{t \in \mathcal{T}_p}, \left(CF_t^L\right)_{t \in \mathcal{T}_p}$, namely one for the assets and another one for the liabilities. We assume that they are given as outputs from the corresponding CFP model. Moreover, as a simplification of notation the resulting profit/loss cash flows, i.e. $CF_t = CF_t^A - CF_t^L$, will be examined. Thus, it is possible to define the respective TV as the sum of all cash flow projections accrued to maturity date T, i.e.

$$\text{TV} := \sum_{t=t_1}^{T} \frac{N_T}{N_t} CF_t^A - \frac{N_T}{N_t} CF_t^L = \sum_{t=t_1}^{T} \frac{N_T}{N_t} CF_t = N_T \sum_{t=t_1}^{T} \widetilde{CF}_t.$$

Here, $\widetilde{CF}_t$ denotes the discounted cash flow which in turn reflects the scenario-wise present value at time t. The numéraire process $(N_t)_{t \in \mathcal{T}}$ allows, furthermore, a derivation of the discounted values at t_1, i.e. the Discounted Terminal Value (DTV) is given by

$$\text{DTV} := \frac{N_{t_1}}{N_T} \cdot \text{TV}. \tag{1.6}$$

Note that, in general, DTV is not $\mathcal{F}_{t_1}$-measurable, but only $\mathcal{F}_T$-measurable, i.e. $\text{DTV} \in L^2(\Omega, \mathcal{F}_T, \mathbb{P})$. Later on, we will see that this is the needed value for determining the market values since it reflects the scenario-wise cash flow values at time $t = t_1$. In this setup we assume that $CF_t \in L^2(\Omega, \mathcal{F}_t, \mathbb{P})$ holds and thus $\text{DTV} \in L^2(\Omega, \mathcal{F}_T, \mathbb{P})$ follows directly.

Market Values & Basic own Funds

In accordance to the Solvency II principles and based on the simplified balance sheet the corresponding distribution of the market values of assets and liabilities must be determined at time t_1. In mathematical terms, the market value of assets less the liabilities at time t_1 is given by the conditional expectation of the Discounted Terminal Value at t_1 conditioned on the market situation at t_1, i.e.,

$$\begin{aligned}
\text{BoF}_{t_1} &= MV_{t_1}^A - MV_{t_1}^L \\
&= \mathbb{E}_{\mathbb{Q}}\left[\sum_{t=t_1}^{T} \frac{N_{t_1}}{N_t} CF_t^A \,\bigg|\, \mathcal{F}_{t_1} \right] - \mathbb{E}_{\mathbb{Q}}\left[\sum_{t=t_1}^{T} \frac{N_{t_1}}{N_t} CF_t^L \,\bigg|\, \mathcal{F}_{t_1} \right] \\
&= \mathbb{E}_{\mathbb{Q}}\left[\sum_{t=t_1}^{T} \frac{N_{t_1}}{N_t} CF_t \,\bigg|\, \mathcal{F}_{t_1} \right] = \mathbb{E}_{\mathbb{Q}}\left[\frac{N_{t_1}}{N_T} \sum_{t=t_1}^{T} \frac{N_T}{N_t} CF_T \,\bigg|\, \mathcal{F}_{t_1} \right] \\
&= \mathbb{E}_{\mathbb{Q}}\left[\text{DTV} \,|\, \mathcal{F}_{t_1} \right]. \tag{1.7}
\end{aligned}$$

Since t_1 is the first period after t_0 it exists according to Natolski (2018) (cf. Theorem 4.3.4) an equivalent risk neutral risk measure $\mathbb{Q}$ such that $\frac{d\mathbb{P}}{d\mathbb{Q}}\big|_{\mathcal{F}_{t_1}} \in L^\infty(\Omega, \mathcal{F}_{t_1}, \mathbb{Q})$ holds. Then, based on the definition of the classical p-norm under the probability measure $\mathbb{P}$ we obtain

$$\left\| \mathbb{E}_{\mathbb{Q}} \left[\text{DTV} | \mathcal{F}_{t_1} \right] \right\|_{L^2(\mathbb{P})} = \sqrt{\mathbb{E}_{\mathbb{P}} \left[\left(\mathbb{E}_{\mathbb{Q}} \left[\text{DTV} | \mathcal{F}_{t_1} \right] \right)^2 \right]}.$$

Applying the conditional Jensen inequality yields

$$\sqrt{\mathbb{E}_{\mathbb{P}} \left[\left(\mathbb{E}_{\mathbb{Q}} \left[\text{DTV} | \mathcal{F}_{t_1} \right] \right)^2 \right]} \le \sqrt{\mathbb{E}_{\mathbb{P}} \left[\mathbb{E}_{\mathbb{Q}} \left[\text{DTV}^2 | \mathcal{F}_{t_1} \right] \right]}$$

$$= \sqrt{\mathbb{E}_{\mathbb{Q}} \left[\frac{d\mathbb{P}}{d\mathbb{Q}} \cdot \mathbb{E}_{\mathbb{Q}} \left[\text{DTV}^2 | \mathcal{F}_{t_1} \right] \right]}.$$

Since $\mathbb{E}_{\mathbb{Q}} \left[\mathbb{E}_{\mathbb{Q}} \left[X \cdot Y | \mathcal{F}_{t_1} \right] \right] = \mathbb{E}_{\mathbb{Q}}[X \cdot Y] = \mathbb{E}_{\mathbb{Q}} \left[\mathbb{E}_{\mathbb{Q}} \left[X | \mathcal{F}_{t_1} \right] \cdot Y \right]$ holds, if Y is $\mathcal{F}_{t_1}$-measurable, we get

$$\mathbb{E}_{\mathbb{Q}} \left[\frac{d\mathbb{P}}{d\mathbb{Q}} \cdot \mathbb{E}_{\mathbb{Q}} \left[\text{DTV}^2 | \mathcal{F}_{t_1} \right] \right] = \mathbb{E}_{\mathbb{Q}} \left[\mathbb{E}_{\mathbb{Q}} \left[\frac{d\mathbb{P}}{d\mathbb{Q}} \cdot \mathbb{E}_{\mathbb{Q}} \left[\text{DTV}^2 | \mathcal{F}_{t_1} \right] \mid \mathcal{F}_{t_1} \right] \right]$$

$$= \mathbb{E}_{\mathbb{Q}} \left[\mathbb{E}_{\mathbb{Q}} \left[\frac{d\mathbb{P}}{d\mathbb{Q}} \mid \mathcal{F}_{t_1} \right] \cdot \mathbb{E}_{\mathbb{Q}} \left[\text{DTV}^2 | \mathcal{F}_{t_1} \right] \right]$$

$$= \mathbb{E}_{\mathbb{Q}} \left[\left. \frac{d\mathbb{P}}{d\mathbb{Q}} \right|_{\mathcal{F}_{t_1}} \cdot \mathbb{E}_{\mathbb{Q}} \left[\text{DTV}^2 | \mathcal{F}_{t_1} \right] \right],$$

with $X = \frac{d\mathbb{P}}{d\mathbb{Q}}$ and $Y = \mathbb{E}_{\mathbb{Q}} \left[\text{DTV}^2 | \mathcal{F}_{t_1} \right]$. Overall, we obtain thus

$$\sqrt{\mathbb{E}_{\mathbb{Q}} \left[\frac{d\mathbb{P}}{d\mathbb{Q}} \cdot \mathbb{E}_{\mathbb{Q}} \left[\text{DTV}^2 | \mathcal{F}_{t_1} \right] \right]} = \sqrt{\mathbb{E}_{\mathbb{Q}} \left[\left. \frac{d\mathbb{P}}{d\mathbb{Q}} \right|_{\mathcal{F}_{t_1}} \cdot \mathbb{E}_{\mathbb{Q}} \left[\text{DTV}^2 | \mathcal{F}_{t_1} \right] \right]}$$

Here, $\text{DTV} \in L^2(\Omega, \mathcal{F}_T, \mathbb{P})$ holds and according to Föllmer and Schied (2016) (Theorem 5.16) $\frac{d\mathbb{Q}}{d\mathbb{P}} \in L^\infty(\Omega, \mathcal{F}_t, \mathbb{P})$. This yields

$$\|\text{DTV}\|_{L^2(\mathbb{Q})} = \sqrt{\mathbb{E}_{\mathbb{Q}} \left[\text{DTV}^2 \right]} = \sqrt{\mathbb{E}_{\mathbb{P}} \left[\frac{d\mathbb{Q}}{d\mathbb{P}} \cdot \text{DTV}^2 \right]}$$

$$\le \left\| \frac{d\mathbb{Q}}{d\mathbb{P}} \right\|_{L^\infty(\mathbb{P})} \cdot \|\text{DTV}\|_{L^2(\mathbb{P})} < \infty,$$

i.e. $\text{DTV} \in L^2(\Omega, \mathcal{F}_T, \mathbb{Q})$. Then, the general Hölder inequality and $\left. \frac{d\mathbb{P}}{d\mathbb{Q}} \right|_{\mathcal{F}_{t_1}} \in L^\infty(\Omega, \mathcal{F}_{t_1}, \mathbb{Q})$ leads finally to

$$\sqrt{\mathbb{E}_{\mathbb{Q}}\left[\left.\frac{d\mathbb{P}}{d\mathbb{Q}}\right|_{\mathcal{F}_{t_1}} \cdot \mathbb{E}_{\mathbb{Q}}\left[\mathrm{DTV}^2 \mid \mathcal{F}_{t_1}\right]\right]} \leq \left\|\left.\frac{d\mathbb{P}}{d\mathbb{Q}}\right|_{\mathcal{F}_{t_1}}\right\|_{L^\infty(\mathbb{Q})} \cdot \|\mathrm{DTV}\|_{L^2(\mathbb{Q})} < \infty,$$

and thus $\mathrm{BoF}_{t_1} \in L^2(\Omega, \mathcal{F}_{t_1}, \mathbb{P})$. Furthermore, an immediate consequence is that the same applies to $MV_{t_1}^A$ and $MV_{t_1}^L$, if $\widetilde{CF}_t^A$ and $\widetilde{CF}_t^L \in L^2(\Omega, \mathcal{F}_t, \mathbb{P})$ for all $t \in \mathcal{T}_p$. Note, that these results can similarly be enhanced to the general $L^p(\mathbb{P})$ case for $p > 1$.

As already mentioned at the beginning, we obtain the SCR as the negative upper $1 - \alpha$-quantile of the BoF_{t_1}, i.e.

$$\mathrm{SCR} := -q_{1-\alpha,+}^{\mathrm{BoF}_{t_1}}, \tag{1.8}$$

with $\alpha = 99.5\%$. Now, with this analytical setup in a next step we consequently introduce the corresponding Monte Carlo simulation framework in order to approximate, first, the conditional expectation (1.7) and, second, the Solvency Capital Requirement (1.8).

1.4   Nested Monte Carlo

Since an analytical derivation of the market value of liabilities is, according to the complex payoff structures, impossible, approximation methods are indispensable. Here, especially Monte Carlo simulations are a logical choice due to the fact that the BoF_{t_1} problem boils down to an approximation of a conditional expectation. Consequently, we briefly introduce the key ideas behind the so-called moment- and quantile-based nested Monte Carlo simulations and link the latter one to the aforementioned actuarial requirements at hand.

Monte Carlo and Nested Monte Carlo in General

In accordance with the strong law of large numbers, Monte Carlo methods gained attention as a reliable asymptotic approximation tool for expected values. In a general manner, the law states that for an underlying sample $\xi_1, \ldots, \xi_n$ for $n \in \mathbb{N}$ with $\xi_i \sim \xi$ iid

$$\frac{1}{n}\sum_{i=1}^{n} \xi_i \xrightarrow[n\to\infty]{a.s.} \mathbb{E}[\xi]$$

holds. The key take away message is that the empirical mean serves as a consistent point estimator for the mean. Now, in order to obtain an estimator to approximate conditional expectations in $L^2(\Omega, \mathcal{F}_{t_1}, \mathbb{P})$, like in setting (1.7), as a logical consequence two Monte Carlo methods have to be concatenated. One to approximate the underlying conditioning space and the other to derive the respective expectation. We will denote this as a nested Monte Carlo simulation.

For insurance specific requirements such a brute force method seems, as summarized in Figure 1.4, to be the obvious choice. Then, we approximate the underlying market values (1.7) at t_1 conditioned at the present state process scenario between t_0 and t_1. In this regard, we simulate at each outer stage the ESG risk scenarios up to time t_1. At each inner stage, however, the asset and liability paths (ALM scenarios) are simulated during the projection horizon (cf. Figure 1.4).

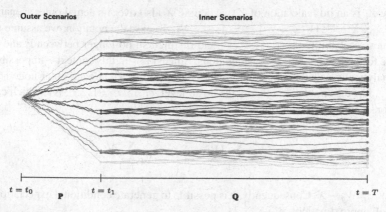

Figure 1.4 Illustration of an exemplary nested simulation tree. The outer scenarios are generated under the real world measure $\mathbb{P}$ and during the risk horizon $t = t_0, \ldots, t_1$. Conditional on each outer node, during the projection horizon from t_1 to maturity T, inner scenarios were generated for evaluation. They are generated under the risk-neutral measure $\mathbb{Q}$

In mathematical terms, we assume a given computational or scenario budget $N := N_1 \cdot N_2$ with $N_1 \in \mathbb{N}$ outer and $N_2 \in \mathbb{N}$ inner scenarios. Let $Z_1, \ldots, Z_{N_1}$ denote the underlying risk factors or ESG scenarios at time t_1. These are iid random variables drawn from $Z = \mathbf{Z}_{t_1}$. Aiming at projecting the assets and liabilities of the insurer into the future, more precisely from t_0 to t_1 (cf. risk horizon), these risk scenarios have to be generated under the real world measure $\mathbb{P}$. Figure 1.4 indicates the real world resp. risk scenarios as outer scenarios (outers). Note, that the real world measure is a physical measure and can thus be obtained from historical data.

According to the market-consistence valuation principle each conditional expectation of the underlying assets or liabilities has to be reevaluated under the respective risk scenario set, cf. BoF (1.3) at t_1. Hence, by $V_{i1}, \ldots, V_{iN_2}, i = 1, \ldots, N_1$, we denote the occurring Discounted Terminal Values at the respective risk scenario Z_i. These in turn are drawn from $V|Z = Z_i$ and are thus conditionally iid (i.e. conditional independent (cid) and identical distributed) in $j = 1, \ldots, N_2$ given Z. Thus, in the following, a consideration of V_j instead of $V_{i1}, \ldots, V_{iN_2}, i = 1, \ldots, N_1$ is sufficient. Note, that according to the definition of (1.7) these scenarios have to be risk neutral and will thus be generated under $\mathbb{Q}$. The respective cdfs of F_Z and $F_{V|Z}$ are generally unknown. Overall, we can denote the Basic own Funds at t_1 by

$$X := \mathbb{E}_{\mathbb{Q}} \left[\text{DTV} | \mathcal{F}_{t_1} \right] = \mathbb{E}_{\mathbb{Q}} \left[V | Z \right]. \tag{1.9}$$

Since Z_i is an iid realization of Z, we denote X also over its actual outer scenario state, i.e. $X_i = \mathbb{E}\left[V | Z = Z_i \right]$, $i = 1, \ldots, N_1$. To ease the notation, we assume an atomless probability space $(\Omega, \mathcal{F}, \mathbb{P})$ and distinguish no longer between $\mathbb{P}$ and $\mathbb{Q}$ in the following theoretical considerations. Since it is—in this context—impossible to analytically calculate the conditional expectation (1.9), at each outer node $i = 1, \ldots, N_1$, the simulation inherits the noise level, ϵ_{i,N_2}, $i = 1, \ldots, N_1$, which can be considered as iid realizations of the random variable

$$\epsilon_{N_2} := \frac{1}{N_2} \sum_{j=1}^{N_2} \left(V_j - X \right) = \frac{1}{N_2} \sum_{j=1}^{N_2} W_j,$$

for $W_j := V_j - X$. Consequently, it is possible to generate conditional expectations or BoF approximations $\bar{X}$, i.e.

$$\bar{X} := X + \epsilon_{N_2}.$$

Now, these definitions give rise to the following consequences:

Proposition 1.4.1
If $V \in L^p (\Omega, \mathcal{F}, \mathbb{P})$ for some $p \geq 1$, then $||W_j||_{L^p(\Omega,\mathcal{F},\mathbb{P})} \leq ||V||_{L^p(\Omega,\mathcal{F},\mathbb{P})}$ for all $j \in \mathbb{N}$ and similarly $||\epsilon_{N_2}||_{L^p(\Omega,\mathcal{F},\mathbb{P})} \leq ||V||_{L^p(\Omega,\mathcal{F},\mathbb{P})}$ for all $N_2 \in \mathbb{N}$. Further, in this case, $\mathbb{E}[W_j \mid Z] = \mathbb{E}[\epsilon_{N_2} \mid Z] = 0$, and especially $\mathbb{E}[W_j] = \mathbb{E}[\epsilon_{N_2}] = 0$. Most importantly, a law of large numbers for $\{W_j\}_{j\in\mathbb{N}}$ holds:

$$\epsilon_{N_2} \xrightarrow[N_2\to\infty]{\text{a.s.}} 0 \quad \text{and} \quad \epsilon_{N_2} \xrightarrow[N_2\to\infty]{L^1} 0.$$

If $p \geq 2$, then it further holds for $i, j \in \mathbb{N}$ that

$$\mathbb{E}[W_j^2 \mid Z] = \mathbb{E}[V^2 \mid Z] - X^2, \quad \mathbb{E}[W_i W_j \mid Z] = 0 \text{ for } i \neq j, \quad \mathbb{E}[\epsilon_{N_2}^2 \mid Z] = \frac{1}{N_2} \mathbb{E}[W_j^2 \mid Z].$$

In addition, for $p \geq 2$ a central limit theorem holds for $\{W_j\}_{j \in \mathbb{N}}$:

$$\frac{\sqrt{N_2}}{\mathbb{E}[V^2 \mid Z] - X^2} \cdot \epsilon_{N_2} \xrightarrow[N_2 \to \infty]{d} N(0, 1).$$

Last but not least, the following (conditional) Rosenthal type inequalities hold (almost surely) for all N_2, depending on the specific value of $p \geq 1$:

$$1 \leq p \leq 2: \quad \mathbb{E}[\,|\epsilon_{N_2}|^p \mid Z] \leq C_p \cdot N_2^{-p+1} \cdot \mathbb{E}[\,|W_j|^p \mid Z],$$
$$p > 2: \quad \mathbb{E}[\,|\epsilon_{N_2}|^p \mid Z] \leq C_p \cdot \left(N_2^{-p+1} \cdot \mathbb{E}[\,|W_j|^p \mid Z] + N_2^{-p/2} \cdot \mathbb{E}[W_j^2 \mid Z]^{p/2} \right),$$

where the constant C_p depends on p only.

Proof.
We start with the following observations,

$$\|\epsilon_{N_2}\|_{L^p} \leq \|W_j\|_{L^p} \leq \|V_j\|_{L^p} + \|X\|_{L^p} \leq 2\|V_j\|_{L^p},$$

which proves the first claim. The second claim follows directly from the fact $X = \mathbb{E}[V_j \mid Z]$. The almost sure law of large numbers for $p \geq 1$ follows for example by Majerak et al. (2005), Theorem 4.2, since the sequence $\{W_j\}_{j \in \mathbb{N}}$ is conditionally iid. Taking into account that the sequence $\{W_j\}_{j \in \mathbb{N}}$ is also cid in the sense of Berti et al. (2004), we also obtain the stated L^1-convergence by Berti et al. (2004), Theorem 2.2. The moment statements for $p \geq 2$ follow from straightforward calculations after conditioning on Z. The central limit theorem for a conditional iid sequence is due to Yuan et al. (2014) Theorem 3.1, noting that $\mathbb{E}[W_j^2 \mid Z] = \mathbb{E}[V^2 \mid Z] - X^2$ has to be used for standardizing. The (conditional) Rosenthal type inequalities follow from Theorem 1 in Yuan et al. (2022), taking into account that a conditionally iid sequence is always conditionally negatively associated. $\square$

Proposition 1.4.1 states a strong law of large numbers for centered random variables $\{W_j\}_{j \in \mathbb{N}}$. According to the missing independence of the W_j classical LLNs are not applicable. Hence, we used the results for conditionally iid sequences, see e.g.

Majerak et al. (2005). But note that for larger p, i.e. $p \geq 2$, classical LLNs apply, since the sequence $\{W_j\}_{j \in \mathbb{N}}$ is, then, uncorrelated.

Moreover, the given central limit theorem relies also on the conditionally iid property of the underlying sequence $\{W_j\}_{j \in \mathbb{N}}$. Yuan et al. (2014) (cf. Theorem 3.1) state an even more refined version on the limiting distribution, which gives further insight into the conditional behaviour of ϵ_{N_2}. Now, based on this simulation framework we introduce two problem statements for future investigations.

First, as intermediate step, the so-called *moment-based* problem will be considered, i.e.

$$\gamma := \mathbb{E}_{\mathbb{P}}\left[G(X)\right], \tag{1.10}$$

for some nonlinear function $G : \mathbb{R} \to \mathbb{R}$. Advanced risk measures like the Value-at-Risk, the Average-Value-at-Risk or the Expected Shortfall are not included in (1.10). Nevertheless, this problem statement formulates somehow the most basic question on loss distributions and gives some important insights for advanced problems. Common examples in literature are the probability of a large loss (LL), i.e. $G(X) = \mathbb{1}_{\{X > c\}}$ for $c \in \mathbb{R}$, or the expected excess loss for the consideration of derivatives in a financial context, i.e. $G(X) = (X - c)^+$ for $c \in \mathbb{R}$. This problem class is rather intuitive because the strong law of large numbers leads obviously to

$$\bar{\gamma}_{N_1, N_2} := \frac{1}{N_1} \sum_{i=1}^{N_1} G\left(\frac{1}{N_2} \sum_{j=1}^{N_2} V_{ij}\right) = \frac{1}{N_1} \sum_{i=1}^{N_1} G\left(X_i + \epsilon_{i, N_2}\right)$$

and thus to an almost sure estimator of γ, see Chapter 3 for more details.

According to the SCR problem formulation, as second, so called *quantile-based* problems are of particular interest, i.e.

$$\rho := -q_\alpha^X, \tag{1.11}$$

for $\alpha \in (0, 1)$, whereby $q_{\alpha,+}^X$ can be used equivalently (cf. SCR definition (1.8) and the discussion therein). Note, that the SCR problem is covered by (1.11) for $\alpha = 0.5\%$. Here, it is compared to the moment-based problem not obvious how to obtain an intuitive almost sure nested Monte Carlo estimator representation. Two approaches are rather popular: Either approximating the quantile function q_α^X by its empirical counterpart over the simulated market values under various risk scenarios or relying on order statistics. On the one hand, according to Serfling (1980) the empirical distribution function of the observations $\bar{X}_1, \ldots, \bar{X}_{N_1}$ is

$$\widehat{F}^{N_1}_{X+\epsilon_{N_2}}(x) = \frac{1}{N_1} \sum_{i=1}^{N_1} \mathbb{1}_{\{X_i + \epsilon_{i,N_2} \leq x\}}, \quad -\infty < x < \infty,$$

i.e. each observation $X_i + \epsilon_{i,N_2}$ receives a mass of $1/N_1$. Then, the quantile, denoted by $\widehat{q}^{X+\epsilon_{N_2}}_{\alpha,N_1}$, of $\widehat{F}^{N_1}_{X+\epsilon_{N_2}}$ serves as estimator of (1.11). This approach attains legitimacy since for non-noisy simulations, i.e. the iid sequence $\{X_i\}$ occurs without error, the Glivenko-Cantelli Theorem holds. It states the uniform almost sure convergence of the empirical distribution function towards the cumulative distribution function (cdf) (cf. Serfling (1980)), i.e.

$$||\widehat{F}^{N_1}_X - F_X||_\infty = \sup_{x \in \mathbb{R}} \left| \widehat{F}^{N_1}_X(x) - F_X(x) \right| \xrightarrow[N_1 \to \infty]{a.s.} 0.$$

Thereby, $||\widehat{F}^{N_1}_X - F_X||_\infty$ describes the Kolmogorov-Smirnov distance and $\widehat{F}^{N_1}_X$ the empirical distribution function for the underlying non-noisy simulation. If additionally

$$\delta(\epsilon, X) := \min \left\{ \alpha - F_X \left(q^X_\alpha - \epsilon \right), F_X \left(q^X_\alpha + \epsilon \right) - \alpha \right\}$$

is strictly positive for all $\epsilon > 0$, the empirical quantile estimator converges towards the analytical quantile almost surely (cf. Serfling (1980) (Theorem, p. 75)). This last condition is a so-called 'non-flatness' condition and ensures that the cdf F_X is strictly increasing, otherwise a convergence of the quantile estimator cannot be achieved. In a noisy nested simulation the underlying noise level ϵ_{i,N_2} plays a crucial role and makes a straightforward application of the Glivenko-Cantelli Theorem impossible since the consideration of the uniform limit over N_1 and N_2 is necessary. Its behaviour depends highly on the underlying computational budget splitting between inner and outer scenarios and has a huge impact on the convergence properties. We will address these issues in more detail in Chapter 4. Nevertheless, it should be mentioned that besides the quantile approximation this approach also provides a cdf approximation of F_X resp. $F_{V|Z}$. This is a nice side effect and in an insurance context also an useful information for risk management tasks. On the other hand, it is possible to use order statistics. Therefore, it is in an interim step necessary to sort, first of all, the underlying Basic own Funds observations $\bar{X}_1, \ldots, \bar{X}_{N_1}$ in ascending order, i.e.

$$\bar{X}_{(1)} \leq \cdots \leq \bar{X}_{(N_1)},$$

then, second of all, $\bar{X}_{(\lfloor N_1 \alpha \rfloor)}$ defines the quantile estimator, cf. David and Nagaraja (2004) and Serfling (1980). The definition states that the quantile is given by the

$\lfloor N_1 \alpha \rfloor$ largest observation, whereby $\lfloor x \rfloor$ describes the largest integer value smaller or equal to x.

Note, that both quantile estimation approaches (by order statistics resp. over the empirical cdf) coincide since the following relation holds

$$\widehat{q}_{\alpha,N_1}^{X+\epsilon_{N_2}} = \begin{cases} \bar{X}_{(N_1\alpha)}, & \text{if } N_1 \cdot \alpha \text{ is an integer,} \\ \bar{X}_{(\lfloor N_1\alpha \rfloor+1)}, & \text{if } N_1 \cdot \alpha \text{ is not an integer,} \end{cases}$$

cf. Serfling (1980) (Section 2.4.1).

Literature Review

The existing literature on nested Monte Carlo simulations can essentially be divided into two categories: On the one hand, different types of convergence and their corresponding convergence rates are treated for the uniform nested Monte Carlo approach. In this introductory part we are focusing on the moment and quantile estimation special case. Note, supplementary results that might be helpful in the course of this work will be referenced in the respective subsequent chapters. On the other hand, modified procedures are introduced to ease the simulation effort, especially due to savings at the inner stage. Most publications support their results with examples from a financial mathematical viewpoint. Hence, insurance-specific peculiarities are only addressed by a few selected publications which will also be considered in an extra section. In the following, we give a brief overview of the current literature state based on the previously mentioned classification.

Standard Nested Monte Carlo and Rates of Convergence for Moment Estimators

For the standard uniform nested Monte Carlo approach only a few publications exist, cf. Table 1.1. Here, especially the L^2 convergence of $\bar{\gamma}_{N_1,N_2}$ towards γ and the convergence in distribution are considered. Obviously, the convergence in distribution follows immediately from the L^2 convergence. In this Section our focus lies only on uniform simulation problems, i.e. $\bar{\gamma}_{N_1,N_2}$. Non-uniform or kernel approaches are briefly summarized in an upcoming section on modified nested Monte Carlo simulations.

Starting with the L^2-convergence, Hong and Juneja (2009) show under quite strong assumptions that the Root Mean Squared Error (RMSE), i.e. $||\bar{\gamma}_{N_1(N),N_2(N)} - \gamma||_{L^2}$, decays asymptotically at a rate of $N^{-1/3}$; a rate that can be obtained for

Table 1.1 Summary of results concerning moment estimator convergence on the basis of a nested Monte Carlo simulation for continuous conditioning spaces

	a.s.	in L^2-norm	in distr.
Convergence	✗	✓	✓
Optimal Rate	✗	$\mathcal{O}\left(N^{-1/3}\right)$	$\mathcal{O}\left(N^{-1/3}\right)$
Authors	✗	(H/J (09)), (G/J (10)), (R (18)), (L (22))	(L (98)), (A/G (16)) (L (22))
Functions	✗	$G \in \mathbf{C}^2$, $G \in \mathbf{C}^3$, LL, HS	LL, (VB,β), $G \in \mathbf{C}^3$, LL, HS

example by a budget split of $N_1 \sim N^{2/3}$ outer and $N_2 \sim N^{1/3}$ inner scenarios. We note that key to this fast decay rate is their observation that under their strong assumptions the *bias* satisfies

$$|\mathbb{E}[G(X + \epsilon_{N_2}) - G(X)]| \leq C \cdot N_2^{-1}$$

i.e. the bias is of order -1 in N_2 (the constant C only depends on the joint distribution of (Z, V) as well as on G). Note, that for their main results, Hong and Juneja (2009) state the following assumptions: X possesses a density $f_X \in \mathbf{C}^1(\mathbb{R})$, $G = \sum_{k=1}^{K} G_k \mathbb{1}_{I_k}$ with $G_k \in \mathbf{C}^3(I_k)$ for suitable disjoint intervals I_k, and $\sup_{x \in \mathbb{R}} |G'''(x)| < \infty$. However, their proof yields that further implicit assumptions concerning the joint density function $f_{(Z,V)}$ of (Z, V) (like differentiability), concerning (conditional) moments of V, and concerning the (uniformity of the) behaviour of the conditional densities $f_{V|Z}$ have to be made to obtain the stated rate. In Gordy and Juneja (2010), these assumptions are summarized and made transparent in a mathematical rigorous way in their Assumption 1. Importantly, in Gordy and Juneja (2010), only the special case $G(x) = \mathbb{1}_I(x)$ is covered (with rate $N^{-1/3}$), while the case of a general function G is no longer covered. A recent publication of Liu et al. (2022) extends the restricted consideration of the large loss problem in Gordy and Juneja (2010) to the cases where $G \in \mathbf{C}^3(\mathbb{R})$ and for hockey-stick functions (HS), i.e. $G(X) = \max\{X, 0\}$. They obtain the same rate $N^{-1/3}$ and use equivalent assumptions as in Gordy and Juneja (2010) (cf. Assumption 1), except the smooth function case. Here, it is sufficient to suppose that $\mathbb{E}\left[G''(X) \cdot V^2\right] < \infty$ holds. The related result of Rainforth et al. (2018), Theorem 3, for $G \in \mathbf{C}^2(\mathbb{R})$ with bounded second derivative, similarly only covers the case of a smooth function G; hence, in our opinion, the case of *piecewise smooth* functions G still remains open for a precise mathematical analysis. The results of Hong and Juneja (2009) and Rainforth et al. (2018) indicate that either the function G has to be quite well-behaved or the joint density of (Z, V) has to be sufficiently smooth to allow for a fast decay of the bias of order -1, which seems to be the key for fast convergence rates in general. Consequently, the application of these results to a risk measure like the probability of a large loss (i.e. $G(X) = \mathbb{1}_{\{X>c\}}$, $c \in \mathbb{R}$) yields an L^2-rate of $N^{-1/3}$ (under smoothness assumption on $f_{(Z,V)}$).

For the convergence in distribution Glynn and Lee (2003) also consider the large loss probability, however, only in the context of a discrete conditioning space, i.e. Z takes only values on a finite or at most countable space. They obtain, based on the budget split $N_1 \sim N/\log(N)$ outer and $N_2 \sim \log(N)$ inner scenarios, convergence in distribution to a normal distribution with a remarkable convergence rate of $(\log(N)/N)^{1/2}$, cf. Glynn and Lee (2003), Theorem 2.2. This result is based on the very strong assumption that $V \mid Z$ possesses a finite (conditional) moment generating function which behaves uniformly with respect to Z. In conjunction with a few additional mild technical assumptions (which hold for almost all $c \in \mathbb{R}$), this yields an exponential decay of the bias in N_2 for the special case of the large loss probability, which in turn yields the above result.

For general conditioning spaces, Andradóttir and Glynn (2016) and Liu et al. (2022) analogously establish related CLT-type convergence rates, see Andradóttir and Glynn (2016), Theorem 3.4 and Liu et al. (2022), Theorem 2. In their setup, Andradóttir and Glynn (2016) do not consider specific function classes for G (nor asymptotic bias rates), but start directly with assuming that a certain asymptotic bias behaviour holds. Liu et al. (2022), however, consider $G \in \mathbf{C}^3(\mathbb{R})$, $G(X) = \max\{X, 0\}$ and $G(X) = \mathbb{1}_{\{X>c\}}$. Here, they need further assumptions: If $G \in \mathbf{C}^3(\mathbb{R})$ holds then $\mathbb{E}\left[|G(X)|^2\right]$, $\mathbb{E}\left[|G'(X)|^2\right]$, $\mathbb{E}\left[|G''(X)|^2\right]$ have to be finite and the fourth moment of V must exist. If $G(X) = \max\{X, 0\}$ (hockey stick case) holds, a finite second moment of V is needed and in the large loss case they use the same assumptions as the already cited assumption of Gordy and Juneja (2010), Assumption 1. Based on these assumptions, optimal rates of order $N^{-1/3}$ for (almost) $N_1 \sim N^{2/3}$ and $N_2 \sim N^{1/3}$ are possible. Note, that we obtain later the same rates for almost sure convergence.

For the special case of the probability of a large loss, the first results can actually be traced back to the PhD thesis of Lee (1998), which seems to lay the basis for the corresponding publication Glynn and Lee (2003). Besides the above mentioned case of a discrete conditioning space (with some additional results for a finite conditioning space), it contains further results in case if X possesses a sufficiently smooth density. The main result is again a bias of order -1 which clearly precedes the same result by Gordy and Juneja (2010), while relying on quite different assumptions (smoothness of f_X, smoothness of the conditional skewness, ...) than Gordy and Juneja (2010).

Standard Nested Monte Carlo and Rates of Convergence for Quantile Estimators

Up to now, the L^2 convergence and convergence in distribution have essentially been investigated for the desired quantile resp. VaR point estimators of q_α^X, cf. Table 1.2.

Gordy and Juneja (2010) rely on order statistics and prove under the already mentioned assumptions that a L^2 convergence holds, with an optimal RMSE rate of $N^{-1/3}$, if $N_1 \sim N^{2/3}$ and $N_2 \sim N^{1/3}$ holds. These findings are underpinned by several portfolio management examples. In addition, they also examine the expected shortfall. This risk measure is an extension of the considered quantile-based problem (1.11) and also important in risk measurement. Gordy and Juneja (2010) obtain an optimal rate of $N^{-1/3}$ under the same assumptions and the same budget split. Furthermore, they show that jackknife methods for bias reduction yield a significant rate improvement and thus $N^{-3/8}$, if $N_1 \sim N^{3/4}$ and $N_2 \sim N^{1/4}$ holds.

Table 1.2 Summary of results concerning quantile estimator convergence on the basis of a nested Monte Carlo simulation

	a.s.	in L²-norm	in distr.
Convergence	✗	✓	✓
Optimal Rate	✗	$\mathcal{O}\left(N^{-1/3}\right)$	$\mathcal{O}\left(N^{-1/3}\right)$
Authors	✗	(G/J (10))	(Lee (98)), (Liu (22))

Again, based on order statistics the remarkable thesis of Lee (1998) (cf. Theorem 3.2.1) and Liu et al. (2022) (cf. Theorem 2) showed that the scaled difference of the nested Monte Carlo quantile estimator and the analytical quantile converge towards a normal distribution and establish thus a CLT-type result. Indeed, this describes the convergence in distribution. In Lee (1998) a distinction between the discrete (finite resp. countably infinite space) and the continuous conditioning space of Z is explicitly made. We are interested in a general setup and thus do not make such distinctions in the following. The discrete part is published in Glynn and Lee (2003), while the continuous part seems to be unpublished so far. Lee (1998) uses the assumptions summarized earlier for the large loss case, whereas Liu et al. (2022) rely on the mentioned assumption set of Gordy and Juneja (2010). Overall, both prove a speed of convergence in distribution towards a normal distribution of $N^{-1/3}$, if (almost) $N_1 \sim N^{2/3}$ and $N_2 \sim N^{1/3}$ holds (cf. Lee (1998), Theorem 3.2.1 and Liu et al. (2022), Theorem 2).

Confidence Intervals for Standard Nested Monte Carlo Approaches

Up to now, we summarized the convergence results for point estimators. Another important statistical perspective is the consideration of confidence intervals (in particular its width) and thus an interval around the quantile q_α^X which covers it with a predefined probability. In this setting asymptotic confidence intervals are investigated because the additional inner simulation requires CLT-like approximations. Hence, we are interested in an $\bar{N}_1 \in \mathbb{N}$ such that

$$\forall N_1 \geq \bar{N}_1 : \quad \lim_{N_2 \to \infty} \mathbb{P}\left(LB_{N_2,\epsilon} \leq q_\alpha^X \leq UB_{N_2,\epsilon}\right) \geq 1 - \epsilon$$

holds, whereby $\epsilon > 0$ describes the predefined confidence level and $LB_{N_2,\epsilon}$ the lower and $UB_{N_2,\epsilon}$ the upper bound of the confidence interval. Hence, q_α^X lies with a probability of at least $1 - \epsilon$ in the interval $[LB_{N_2,\epsilon}, UB_{N_2,\epsilon}]$. For nested Monte Carlo simulations Lan et al. (2007a, b, 2010) investigated several confidence region methods for different risk measures. In Lan et al. (2007b) they derived intervals for the quantile resp. the VaR, in Lan et al. (2007a) for the tail conditional expectation (TCE) and finally in Lan et al. (2010) for the expected shortfall/AVaR (Average Value at Risk). The latter two approaches are relying on empirical likelihood, whereas the here considered VaR case uses well-known central limit (CLT) properties in combination with the fact that for a non nested Monte Carlo approach with exactly known $X_1, \ldots, X_{N_1}$ observations

$$\mathbb{P}\left(X_{(r)} \leq q_\alpha^X \leq X_{(s)}\right) \geq \sum_{j=r}^{s-1} \binom{N_1}{j} \alpha^j (1-\alpha)^{N_1-j} \tag{1.12}$$

holds, cf. David and Nagaraja (2004).

Lan et al. (2007b) assume mainly two error sources ϵ_{out}, ϵ_{in} such that the total coverage probability $\epsilon = \epsilon_{out} + \epsilon_{in} \in (0, 1)$. The outer failure ϵ_{out} emerges due to an unfortunate sampling of risk factors, the inner failure ϵ_{in} on the other hand results from an unlucky sampling of inner level scenarios. In a first step, Lan et al. (2007b) obtain, based on the Student-t distribution, asymptotic confidence regions around each X_i, i.e.

$$\forall i = 1, \ldots, N_1 : \quad \lim_{N_2 \to \infty} \mathbb{P}\left(\bar{X}_i - \rho_{i,N_2} \leq X_i \leq \bar{X}_i + \rho_{i,N_2}\right) \geq 1 - \kappa,$$

with $\rho_{i,N_2} = t_{N_2-1,1-\frac{\delta_L}{2}} \cdot \frac{\hat{\sigma}_i}{\sqrt{N_2}}$. Here, $\delta_L = 1 - (1 - \epsilon_{in})^{\frac{1}{N_1}}$ and $t_{k,l}$ denotes the quantile of a Student-t distribution at the $1 - \frac{l}{2}$-level with $N_k - 1$ degrees of freedom. Based on this κ they obtain (by independence of outer scenarios) by the cartesian product over all intervals a confidence region which captures the emerging inner error with a coverage probability of at least $1 - \epsilon_{in}$. In a second step, to capture the outer stage simulation error they use the common fact (1.12). For the predefined target coverage probability of $1 - \epsilon_{out}$ this in fact yields indices $s_{\epsilon_{out}}$ and $r_{\epsilon_{out}}$. Finally, these indices are used to derive the lower respective upper bound

$$LB_{N_2,\epsilon_{out}} = (\bar{X} - \rho_{N_2})_{(r_{\epsilon_{out}})} \quad \text{and} \quad UB_{N_2,\epsilon_{out}} = (\bar{X} + \rho_{N_2})_{(s_{\epsilon_{out}})}.$$

Note, that $(\bar{X} \pm \rho_{N_2})_{(k)}$ denotes the k-th order statistic from $\{\bar{X}_i \pm \rho_{i,N_2}\}_{i=1}^{N_1}$. Thus, Lan et al. (2007b) obtain an asymptotic $(1 - \epsilon)$-confidence interval for quantiles and in combination with $VaR_\alpha(X) = -q_\alpha^X$ also for the Value-at-Risk. Additionally, the authors briefly discuss the advantages and disadvantages of some efficiency enhancement tools, like e.g. the possibility of screening out undesirable inner scenarios.

Based on an order statistic estimator and on their CLT-type result, i.e.

$$\sqrt{N_1}\left(\hat{q}_\alpha^X - q_\alpha^X\right) \Rightarrow \mathcal{N}(0, \sigma^2),$$

Liu et al. (2022) (Theorem 2) also establish a confidence interval procedure with $\sigma^2 = \alpha(1 - \alpha)/f_X^2(q_\alpha^X)$. It is quite obvious that an asymptotic confidence interval immediately follows if σ^2 can be approximated by a respective estimator $\hat{\sigma}_{N_1,N_2}^2$. Hence, Liu et al. (2022) introduce an adequate estimator and show, in their Proposition 1, that under the same assumptions as in Theorem 2, $\hat{\sigma}_{N_1,N_2}^2 \to \sigma^2$ holds in probability for $N_1, N_2 \to \infty$. Hence, for $\epsilon > 0$ a $100 \cdot (1 - \epsilon)$ valid confidence interval is given by

$$\left(\hat{q}_\alpha^X - \frac{z_{1-\epsilon/2} \cdot \hat{\sigma}_{N_1,N_2}}{\sqrt{N_1}}, \, \hat{q}_\alpha^X + \frac{z_{1-\epsilon/2} \cdot \hat{\sigma}_{N_1,N_2}}{\sqrt{N_1}}\right),$$

whereby $z_{1-\frac{\epsilon}{2}}$ denotes the $1 - \epsilon/2$ quantile of the standard normal distribution. Since the CLT applies also for the moment-based problem, we can replace $\hat{q}_\alpha^X, q_\alpha^X$ also by $\bar{\gamma}_{N_1,N_2}, \gamma$, whereby here $\sigma^2 = Var(G(X))$ holds. Then, obviously, a respective estimator $\hat{\sigma}_{N_1,N_2}^2$, cf. Proposition 1 and previous remarks, can be found and thus also a valid confidence interval.

It should be noted that for nested Monte Carlo estimators no explicit theoretical results concerning the convergence in probability are existing up to now.

Standard Nested Monte Carlo in the Actuarial Context

For the Solvency II risk management case at hand the results of Gordy and Juneja (2010) and Lan et al. (2007b) were adopted and applied by Bauer et al. (2010) resp. in the corresponding dissertation of Bergmann (2011). In this respect, they adapt the existing theory and embed it into the actuarial framework. Hence, the deterministic Basic own Funds at time t_0 has to be considered additionally (cf. (1.2), (1.5)). Beside this adjustment several extensive and practice-oriented numerical analyses round off the publication. Furthermore, Kalberer (2012a, b) summarizes the uniform SCR nested simulation framework briefly and discusses some common challenges (like e.g. the computing time, estimation error, etc.) under strong normality assumptions which are rarely meet in practice. In order to reduce the computing time of the underlying nested simulation Kalberer (2012b) introduces, similar as Broadie et al. (2011) for moment-based problems, a non-uniform approach for quantile-based problems.

Modified Nested Monte Carlo Procedures

The addressed uniform nested simulations are computationally burdensome according to the additional N_1 inner simulations with N_2 scenarios. Based on this downside practitioners struggle especially with the performance of such brute force methods. Hence, some approaches evolved over time to overcome this curse of the uniform nested procedure. Unfortunately, these methods are mostly restricted to the simpler case of approximating $\mathbb{E}[G(X)]$, i.e. the moment-based case. This in turn does not cover the insurance relevant case of a risk measure but nevertheless opens some insights for practitioners to potential computational effort reductions.

Hong and Juneja (2009) propose besides the already seen uniform investigations on moment-based problems also a modified nested Monte Carlo simulation based on a kernel method estimation. Therefore, they assume that G is sufficiently smooth everywhere and use the standard Nadaraya-Watson kernel estimator as approximation of X for the observations $\{(Z_i, V_i)|1 \leq i \leq n\}$, i.e.

$$\tilde{m}_n(x) = \frac{\sum_{i=1}^{n} V_i \, K_h(x - X_i)}{\sum_{i=1}^{n} K_h(x - X_i)},$$

with $K_h(x) = (1/h^d)K(x/h)$, $d \geq 1$, bandwidth h and kernel $K(x) = \mathbb{1}_{\{|x|<1\}}$. The budget split of $N_1 = N$ outers and $N_2 = 1$ inners leads, indeed, to a remarkable RMSE rate of convergence of $N^{-\min\{1/2, 2/(d+2)\}}$, whereby d denotes the dimension of the conditioning space and thus of the random variable Z. Hence, for $d < 4$ faster convergence rates compared to the uniform setting are possible, cf. for example Gordy and Juneja (2010) or Liu et al. (2022). Since $d = 17$ risk factors are not uncommon in insurance related applications, cf. Krah et al. (2018), the application scope of such methods will be tremendously restricted. Hong et al. (2017) repeat the already summarized result (also the rates of convergence) for the d-dimensional uniform kernel $K(x) = \prod_{i=1}^{d} \mathbb{1}_{\{-1/2 \leq x_i \leq 1/2\}}$ for $x = (x_1, \ldots, x_d)$ and enhance the function class by the nondifferentiable $G(X) = \max \{X, 0\}$ (hockey stick function) and discontinuous case $G(X) = \mathbb{1}_{\{X > c\}}$ (i.e. the large loss case).

Broadie et al. (2011) propose a non uniform nested Monte Carlo simulation for the large loss problem ($G(x) = \mathbb{1}_{\{x > c\}}$), i.e. depending on the i-th outer scenario a variable number of $N_{i,2}$, $i = 1, \ldots, N_1$ inners will be generated. The main idea is to add—in a sequential fashion—inner scenarios to loss approximations which are closely related to the threshold level $c \in \mathbb{R}$ of the large loss probability. Then, for losses close to the threshold a distinction is rather difficult. Thus, a variance reduction according to a higher number of inner scenarios leads to an even more precise loss approximation and allows, indeed, a clearer categorization. Consequently, this approach improves the RMSE rate of convergence, according to a variable number of inners, to $N^{-2/5+\epsilon}$, for an arbitrary $\epsilon > 0$.

Feng and Peng (2021) postulate a so-called sample recycling method to reduce the number of inner scenarios. This approach reuses simulated inner scenarios for various outer scenarios. Hence, different scenario sets (an outer one with a normally non matching inner set) will be concatenated by using the respective Radon-Nikodym derivative as weighting factor in the corresponding Monte Carlo estimator. Note, that their analysis concentrates on the large loss risk measure (i.e. $G(X) = \mathbb{1}_{\{X > c\}}$) and that an analysis of the underlying convergence rate is missing. Thus, there is no guarantee that the underlying estimator converges towards the analytical solution. Only a brief comparison between the needed computational budget of the standard and the sample recycling method was carried out.

Besides their analysis on the uniform estimator Liu et al. (2022) establish also a bootstrap method to ease the computational burden in the simulation framework. They draw bootstrap samples $\{Z_1, \ldots, Z_{\bar{N}_1}\}$ and $\{V_{i1}, \ldots, V_{i\bar{N}_2}\}$, $i = 1, \ldots, \bar{N}_1$ and $\bar{N}_1 < N_1$, $\bar{N}_2 < N_2$, i.e. independently and randomly drawn samples with replacement, from an initial sample $\{Z_1, \ldots, Z_{N_1}\}$, $\{V_{i1}, \ldots, V_{iN_2}\}$. In order to obtain a RMSE rate of convergence a minimum sample size of at least $N_1 = N_2^{5/3+\delta}$ for some $\delta > 0$ is needed. Since the computational budget reduces according to the bootstrap sampling to $\bar{N} = \bar{N}_1 \cdot \bar{N}_2$, i.e. $\bar{N} < N$ holds, this leads to an optimal RMSE rate of convergence of $\bar{N}^{-1/3}$ (see Liu et al. (2022), Theorem 1) for

$G \in \mathbf{C}^3(\mathbb{R})$, $G(X) = \max\{X, 0\}$, $G(X) = \mathbb{1}_{\{X>c\}}$ and the quantile problem (1.11). Note that the necessary assumptions are already stated in the previous consideration of the uniform estimator.

1.5   Alternative Monte Carlo Approaches

With the objective to reduce the simulation effort in the approximation of (1.7) two well-known and frequently applied methods in risk management evolved, the so-called Least Squares Monte Carlo (LSMC) and Replication approach. The main concept is the combination of standard regression with Monte Carlo methods grounding on a $L^2(\Omega, \mathcal{F}_{t_1}, \mathbb{P})$ assumption on X. This, indeed, leads to a significant reduction in the inner scenario simulation and thus an approximation of the SCR can be achieved, with the benefit of requiring noticeable less computer capacity. The two existing regression possibilities will be addressed in the following.

Least Squares Monte Carlo (LSMC)

This approach originates from the American or Bermudan option pricing problem and was introduced resp. intensively investigated by Carriere (1996), Longstaff and Schwartz (2001), Tsitsiklis and van Roy (2021), Glasserman and Yu (2004), Egloff et al. (2007) and Zanger (2018) in order to find lower bounds of the option price. Importantly, Clément et al. (2002) show that the approach converges if various assumptions on the basis functions hold. In contrast to the simpler case of an European option, the additional possibility of at least several exercising times creates major difficulties. Each evolution of the stock price and the associated calculation of conditional expectations at each single time point has to be considered. Trying to tackle this problem with a nested simulation procedure is—especially for a huge basket of underlying assets—infeasible because the number of scenarios rises exponentially with its time points. This burden can be avoided by using least squares regression and simulating only unbranched resp. several paths. Overall, this results in an approximation of each conditional expectation resp. the continuation values, by a combination of simple basis functions, such as ordinary polynomials, Chebyshev-, Laguerre- polynomials etc., in every time step.

Insurers, on the other hand, are facing an analogous problem as banking institutions in the before mentioned American option pricing problem. As seen, their risk management requires a calculation of the market value of liabilities in one year (1.7) which boils down to a conditional expectation approximation. Hence, for each real world risk factor scenario we simulate only a few inner projection scenarios. Based on these, a linear regression for the fair value of liabilities will be trained. Thereby, simple basis functions of the risk factors will be used for approximation. The advantage of this setting is that only at time point t_1 a least squares regression has to be carried out and not at every possible time point, as for American options. But it should be noted that this procedure inherits unfortunately a non vanishing projection error because the underlying cash flows for the market value evaluation must be projected across the time interval $[t_1, T]$ for the regression. Since the regression takes place at 'present', the LSMC approach is also well-known as the Regress-Now method, cf. Glasserman and Yu (2004). The whole simulation framework is summarized and depicted in Figure 1.5. Exemplary, Broadie et al. (2015) introduce such a regression based nested Monte Carlo approach for the large loss ($G(X) = \mathbb{1}_{\{X > c\}}$) and two additional risk measures (expected excess loss and squared tracking error, cf. Broadie et al. (2015)). Grounding on given basis functions they solve a classical least squares problem to approximate the BoF estimator. This leads to a remarkable convergence speed improvement of the RMSE given by $N^{-1/2}$. Further theoretical foundations and applications of the LSMC method for evaluating especially life insurance policies can be found in Andreatta and Corradin (2023), Baione et al. (2006), Bernard and Lemieux (2008), Bacinello et al. (2009), Bacinello et al. (2010), Bauer et al. (2012), Krah et al. (2018) and Bauer and Ha (2022). An extension from the standard least squares regression to various

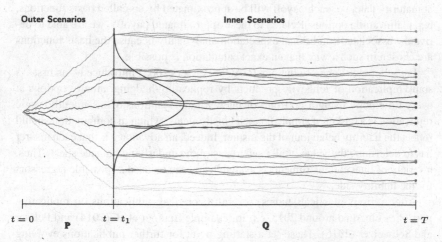

Figure 1.5 Least Squares Monte Carlo approach. The terminal payoffs are generated up to the underlying maturity date T. A linear regression takes place at t_1

adaptive machine learning approaches, like for instance generalized linear models or multivariate adaptive regression splines, can be found in a very recent and application-related publication of Krah et al. (2020b).

Replication

As highlighted in the previous section, the combination of Monte Carlo methods with classical least squares regression results in a significant reduction of inner scenarios. Thus, the least squares Monte Carlo or Regress-Now method leads in comparison to the nested simulations approach to a huge saving of simulation and computing time. Unfortunately, this reduction could only be achieved by accepting a non vanishing projection error according to the BoF regression at t_1. To avoid this clear downside, a new approach, called Regress-Later resp. Replicating portfolios, evolved a few years after the groundbreaking publications on LSMC by Carriere (1996), Longstaff and Schwartz (2001) and Tsitsiklis and van Roy (2021). The pioneering idea lies in approximating the terminal values at the maturity date (i.e. the accrued and aggregated cash-flows) or the cash flow values at each time point instead of approximating the conditional expectation directly (cf. LSMC).

The Regress-Later approach for American and Bermudan option pricing problems was considered firstly from Glasserman and Yu (2004) and Broadie and Cao (2008). As already mentioned, the key idea is an approximation of complex payoffs at maturity date, i.e. each payoff will be approximated by so-called basis functions, e.g. plain vanilla options. Related to these approximated payoffs, we are able to derive the associated conditional expectation analytically because the basis functions are chosen in such a way that an exact calculation is possible.

For guaranteed annuity options (GAOs) Pelsser (2003) introduced—as first—a static replication of longevity products by replicating/hedging annuity options at every time point using a portfolio of interest rate swaptions. Consequently, static replication is a basket of simple financial instruments, which in combination should reflect the liability behaviour of the insurer. Indeed, an advantage is that each insurer has to determine the value of all used assets of the underlying balance sheet. Thus, a variety of financial instruments is available and can be used as possible regressors for the liability side.

After various practical studies, several theoretical publications on replicating portfolios emerged around 2015, c.f. for example Beutner et al. (2014) and Pelsser and Schweizer (2016). These set a starting point for further publications evolving in the upcoming years. Furthermore, they brought this issue in particular in connection with open questions evolving from the actuarial scientific and especially risk management community. Pelsser and Schweizer (2016) for example show that the Regress-Later method from Glasserman and Yu (2004) corresponds to the replicating portfolios approach or in short, the replication. To obtain optimal rates in the replicating portfolio approach the fourth moment of the approximation error must decrease with a rate of $K^{-\gamma}$, $\gamma > 0$. Thereby, K denotes the number of used basis functions / plain vanilla options. Then, it exists a sequence $\mathcal{K} : \mathbb{N} \to \mathbb{N}$ such that the RMSE converges at a rate of $\mathcal{K}(N)^{-\gamma/2}$, cf. Beutner et al. (2014) (Theorem 3.1) and Schweizer (2016) (Theorem 3.1). Hence, if the underlying approximation error vanishes fast enough, rates faster than $N^{-1/2}$ are possible. This is an incredible advantage compared to the LSMC method since the RMSE rate can not exceed $N^{-1/2}$ here.

Figure 1.6 indicates the general and well-known replication approach (in particular the terminal value matching): Each terminal payoff must be replicated, i.e. a linear regression approximates the underlying terminal liability payoffs of the insurer at the maturity date T. The conditional expectations resp. the BoF at t_1 for every underlying real world scenario can be calculated analytically or fast due to the fact of an approximation/ a regression by simple financial products at the end of period. The current state of research and applied practice in actuarial science states two different regression types for replicating portfolios and tackles the construction problem of the searched portfolio. The first approach, called cash flow matching (comparable to the static replicating approach by Pelsser (2003)), regresses at each time point a liability cash flow evolves, i.e. every liability cash flow

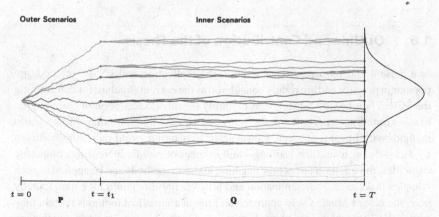

Figure 1.6 Replicating portfolios (terminal value matching). The terminal payoffs are generated up to the underlying maturity date T. The regression takes place at the maturity date based on a portfolio of basic financial instruments

should be approximated as close as possible. The second approach, called terminal value matching (cf. Figure 1.6), introduced by Oechslin et al. (2007), aims at replicating the final terminal payoffs at maturity, i.e. the accrued cash-flows. Indeed, the latter implies a greater flexibility for the insurance company because the cash-flows from t_1 until $T - 1$ can differ and thus for example losses at the beginning can be compensated by upcoming profits. However, it should be mentioned and warned, that two different portfolios, whose terminal values are perfectly matching, can lead to completely different cash-flow payment streams. But luckily, as requested, they lead to exactly the same fair value. A comparison of these construction methods, theoretical foundations and extensions can be found in Natolski and Werner (2014, 2015) and Natolski (2018). In Natolski and Werner (2014) the authors indicate that both methods, i.e. the terminal value and the cash flow matching, lead to the same fair value of the liabilities.

The replicating portfolio method contains in contrast to LSMC only an approximation error. As before, this one is caused by approximating the terminal payoff with a finite instead of an infinite number of regressors and vanishes thus asymptotically with a rising number of used regressors. Note, that this is a clear advantage in contrast to the before mentioned LSMC method. Practical reviews and the usage of replicating portfolios in an insurance liability modeling framework can be found in Daul and Vidal (2009), Chen and Skoglund (2012) and Kalberer (2012a). A fully comprehensive and excellent compendium, which covers the major tasks of replicating life insurance liabilities exemplified for the German life insurance market, can be found in the practical compendium of Seemann (2009).

1.6   Outline and Contribution of the Thesis

In a present and retrospective view both, the replication and LSMC method, are popular in practice and are rightly considered as the current standard for determining the SCR. Their biggest advantage is certainly the drastic reduction of computation time—grounding on regressing the inner scenarios—by combining Monte Carlo methods with linear regression. However, in a changing world increasingly driven by data—through machine learning—and in times of rapidly increasing computing capacities, the question arises how machine learning methods can be applied to such complex tasks as SCR determination and how verifiability could look like. Exactly here, the nested Monte Carlo approach and the mathematical methods therein play a crucial role for the emerging applications and future business models. On the one hand, due to the increased computational capacities, such methods can be applied more frequently. On the other hand, such simulations serve—due to the possible and very important calibration of outer and inner noise levels—as reliable reference value for in- and out-of-sample datasets. Therefore, the remainder of this thesis focuses on nested simulations and their application in the life insurance sector, specializing on the SCR determination. The intended goal is to derive an academical mathematical model that meets the main regulatory Solvency II requirements and to close existing gaps.

Outline

The rest of this thesis is structured as follows:

- Chapter 2 provides a quick overview over the important connection between the almost sure and complete convergence of random variables. Furthermore, relevant probability inequalities and limit theorems will be repeated and adjusted for upcoming investigations and problems at hand.
- In Chapter 3 we discuss the almost sure convergence of nested Monte Carlo estimators for moment-based problem formulations and thus for (1.10). By assuming a moment assumption on $G(X + \epsilon_{N_2})$ we can show the consistency and also an optimal rate of convergence for several function classes G (like for instance bounded, polynomial bounded, Lipschitz continuous functions, ...) with its corresponding minimum growth conditions on N_1 and N_2.
- After this, we turn in Chapter 4 to the almost sure consistency of the nested Monte Carlo quantile estimator. Hence, we will address questions in order to achieve reliable estimates and cover the regulatory frameworks. Therefore, it will be important to generate a sufficiently large set of scenarios, i.e. outer and inner scenarios. But what is an appropriate simulation budget choice, which yields a reliable estimation of the loss distribution? And how should it be divided between outer and inner scenarios? This in turn, is—up to now—for the almost sure convergence (cf. Table 1.2) an open and rather challenging question. We will see that it depends crucially on the given assumptions and also on the complexity of the CFP model. The stated objective of this thesis is to explore some of these issues and answer them in detail from a mathematical point of view with as few assumptions as possible and neglect also company-specific problems (like the present complexity of the CFP model).
- Chapter 5 then turns to the addressed confidence interval problem and introduces, in comparison to Lan et al. (2007b), a new approach for the derivation of such intervals for quantile estimators in a nested simulation environment. For non nested simulations such methods are well known, but fail unfortunately clearly

in a noisy environment, i.e. with a small amount of inner scenarios N_2. Due to this, a separate confidence interval method for nested problems is indispensable. Our approach here relies in general on order statistic theory in combination with binomial distribution and basic CLT properties and culminates in an asymptotic confidence interval around q_α^X for X given by (1.9).

- The practicability of our methods—in an insurance setup—will be demonstrated in Chapter 6. Here, we use our theoretical insights from the previous chapters on quantile estimators and apply them, based on an academic ALM, to the SCR problem (1.5) at hand.
- Finally, Chapter 7 concludes with remarks and possible further research questions. The present and still emerging trend to apply data driven machine learning methods in the risk management of life insurers makes it essential to take a close look at nested Monte Carlo methods, as these turn out to be the only reliable comparison method for such models. This is a crucial point and will explicitly pointed out in more detail in the last chapter and thus gives an outlook for the vibrant machine learning community in risk assessment setups.

Contribution

Grounding on this outline we were able to add the following contributions for moment- and quantile-based estimators in an uniform nested Monte Carlo simulation framework:

- In Chapter 3 we introduce, as first, an uniform almost sure convergence result for moment-based problem formulations and thus enhance the already existing result of Rainforth et al. (2018) (cf. Theorem 2) significantly. Then, the consideration of an uniform limit over $N = N_1 \cdot N_2$ allows a derivation of explicit sequences for inner and outer scenarios which guarantee an almost sure convergence. First of all, we obtain convergence in probability according to the Fuk-Nagaev inequality. Second of all, with the complete convergence and its direct consequence on almost sure convergence we prove under stronger assumptions the strong consistency and derive also an optimal rate of convergence for a polynomial splitting criteria, i.e. $N_1 = N^r$, $N_2 = N^{1-r}$ for $r \in (0, 1)$. Table 1.1 indicates that this convergence type was a missing puzzle piece in the theory on nested Monte Carlo simulations for moment-based problems.
- The novelty of Chapter 4 is that we close the existing gap and provide a proof for the still missing strong consistency (cf. Table 1.2) of the quantile estimator and its corresponding optimal rate. Here, our declared aim was, first of all, to

use as few assumptions as possible and thus to introduce a generally valid statement. In this respect, we were able to waive most of the assumptions of e.g. Lee (1998) in distribution and Gordy and Juneja (2010) in L^2 resp. mean to prove an optimal speed of convergence. Second of all, we investigate the interplay that the usage of further assumptions from the comparable literature leads to better rates. Furthermore, we address extensively the underlying splitting problem of the computational budget N into its single components N_1 (outers) and N_2 (inners) in an almost sure sense. Along the way—by proving the almost sure convergence based on the complete convergence—we also obtain convergence results in probability, cf. Table 1.2. Indeed, these convergence results already follow from the L^2 convergence.

- In Chapter 5 our main contribution is a—to the best of our knowledge—newly developed asymptotic confidence interval method for quantile estimators based on noisy nested Monte Carlo approximations. Our approach in building single intervals around each outer approximation $\bar{X}_i$, $i = 1, \ldots, N_1$ instead of building a huge box around all single intervals (cf. Lan et al. (2007b)) makes a difference for the particular quantile case and simplifies the model parametrisation significantly. This enables, furthermore, a so far novel asymptotic error approximation proof based on the Berry-Esseen Theorem.

- Finally, Chapter 6 applies our theoretical approximation findings to the stated moment- (1.10) and quantile-based (1.11) problems at hand and shows its usability in practice. Hence, our so far mathematical view point will be extended by a numerical analysis based on academic test cases and furthermore by a life insurance-specific example.

Basic Concepts, Probability Inequalities and Limit Theorems 2

Let us, for later considerations, first introduce some common but important concepts of random variables (r.v.) and a corresponding limit theorem. First, explicit upper bounds on the deviation probability between r.v. averages, like $\bar{\gamma}_{N_1,N_2}$ resp. ϵ_{N_2}, and its corresponding mean are needed to establish in Chapter 3 and Chapter 4 the almost sure convergence results. Second, in order to obtain rates for our non parametric confidence interval approach (cf. Chapter 5) the classical Berry-Esseen Theorem will be repeated.

2.1 Complete Convergence

Since we are interested in the asymptotic behaviour of the estimator $\bar{\gamma}_{N_1,N_2}$ in the moment- and $\widehat{q}_{\alpha,N_1}^{X+\epsilon_{N_2}}$ in the quantile-based case, when an available budget of $N \in \mathbb{N}$ is split into $N_1(N) \in \mathbb{N}$ outer and $N_2(N) \in \mathbb{N}$ inner scenarios, we introduce the concept of so-called *budget sequences* for this purpose, i.e.:

Definition 2.1.1.
Here, $\{(N_1(N), N_2(N))\}_{N \in \mathbb{N}}$ denotes a budget sequence, if the following holds:

(i) $N_1(N) \to \infty$, $N_2(N) \to \infty$ for $N \to \infty$,
(ii) $N \ge N_1(N) \cdot N_2(N) \ge N - O(N^{\alpha})$ for some $0 < \alpha < 1$.

© The Author(s), under exclusive license to Springer Fachmedien Wiesbaden GmbH, part of Springer Nature 2024
M. Klein, *Nested Simulations: Theory and Application*, Mathematische Optimierung und Wirtschaftsmathematik | Mathematical Optimization and Economathematics, https://doi.org/10.1007/978-3-658-43853-1_2

A typical example is $N_1(N) = \lfloor N^r \rfloor$, $N_2(N) = \lfloor N^{1-r} \rfloor$ for some $0 < r < 1$. Consequently, a given budget sequence allows to analyze the asymptotic behaviour of the corresponding estimator $\bar{\gamma}_N := \bar{\gamma}_{N_1(N), N_2(N)}$ and $\widehat{\rho}_N := \widehat{q}_{\alpha, N_1(N)}^{X + \epsilon_{N_2(N)}}$.

Introduced by Hsu and Robins (1947) we investigate in following analyses the *complete convergence* of random variable sequences and define it generally as follows:

Definition 2.1.2. *(Hsu and Robins (1947), Theorem 1 resp. Gut (1985))*
A sequence of random variables $\{Y_N\}_{N \in \mathbb{N}}$ *converges completely to a random variable* Y, *i.e.* $Y_N \xrightarrow[N \to \infty]{c} Y$, *if*

$$\forall \epsilon > 0 : \quad \sum_{N=1}^{\infty} \mathbb{P}(|Y_N - Y| > \epsilon) < \infty.$$

According to the well known Borel-Cantelli lemma the desired almost sure convergence is readily implied by complete convergence. Therefore, our main proof strategy to obtain the almost sure convergence is to verify the complete convergence.

Theorem 2.1.3.
Let $\{Y_N\}_{N \in \mathbb{N}}$ *be a sequence of random variables. Then, the complete convergence implies the almost sure convergence immediately, i.e.*

$$Y_N \xrightarrow[N \to \infty]{c} Y \quad \Longrightarrow \quad Y_N \xrightarrow[N \to \infty]{a.s.} Y.$$

Hence, in the following chapters we obtain the almost sure convergence of $\left| \bar{\gamma}_{N_1(N), N_2(N)} - \gamma \right| \xrightarrow[N \to \infty]{a.s.} 0$ and $\left| \widehat{q}_{\alpha, N_1(N)}^{X + \epsilon_{N_2(N)}} - q_\alpha^X \right| \xrightarrow[N \to \infty]{a.s.} 0$ by showing the complete convergence, i.e. $\left| \bar{\gamma}_{N_1(N), N_2(N)} - \gamma \right| \xrightarrow[N \to \infty]{c} 0$ resp. $\left| \widehat{q}_{\alpha, N_1(N)}^{X + \epsilon_{N_2(N)}} - q_\alpha^X \right| \xrightarrow[N \to \infty]{c} 0$.

2.2 Fuk-Nagaev Inequality

Let us recall the Fuk-Nagaev inequality which provides explicit bounds in N and t for the deviation probability of an average of r.v. from the corresponding mean:

Theorem 2.2.1. *(Nagaev (1979), Corollary 1.7)*

Let $p \geq 2$ and $Y_1, \ldots, Y_N$ *be independent random variables with* $\mathbb{E}[Y_i] = 0$ *and* $\mathbb{E}[|Y_i|^p] < \infty$, $i = 1, \ldots, N$. *Then it holds:*

$$\forall t > 0: \quad \mathbb{P}\left(\frac{1}{N}\sum_{i=1}^{N} Y_i \geq t\right) \leq \sum_{i=1}^{N} \mathbb{P}(Y_i \geq \eta t N) + (\eta t N)^{-p} \sum_{i=1}^{N} \mathbb{E}[Y_i^p \cdot \mathbb{1}_{\{0 \leq Y_i \leq \eta t N\}}]$$

$$+ \exp\left(-\frac{(tN)^2}{c_p \cdot \sum_{i=1}^{N} \mathbb{E}[Y_i^2]}\right), \tag{2.1}$$

with $\eta := \frac{p}{p+2}$ *and* $c_p := \frac{e^p (p+2)^2}{p}$.

In our setup, we do not need the Fuk-Nagaev inequality in its basic form (2.1), but in the slightly adjusted formulations (2.2) and (2.3). As we could not find any suitable reference, we provide here explicit proofs for these two immediate consequences of the Fuk-Nagaev inequality.

Corollary 2.2.2.

Let $p \geq 2$ and $\{Y_N\}_{N \in \mathbb{N}}$ *be a sequence of independent and identically distributed random variables with* $\mathbb{E}[Y_1] = 0$ *and* $m_p = \mathbb{E}[|Y_1|^p] < \infty$. *Then, for each* $t > 0$ *there exists some* $\bar{N}(t)$ *such that:*

$$\forall N \geq \bar{N}(t): \quad \mathbb{P}\left(\frac{1}{N}\sum_{i=1}^{N} Y_i \geq t\right) \leq 3 \cdot C_p \cdot t^{-p} \cdot N^{-p+1}. \tag{2.2}$$

Here, $C_p := m_p \cdot \eta^{-p}$ *and* $\bar{N}(t)$ *can be chosen as*

$$\bar{N}(t) := \max\left(\frac{|\ln(C_p) - p\ln(t)|}{p-1}, \frac{4(p-1)^2(c_p m_2)^2}{t^4}\right).$$

Proof.

We apply the Fuk-Nagaev inequality (2.1) and obtain for each N and given $t > 0$:

$$\mathbb{P}\left(\frac{1}{N}\sum_{i=1}^{N} Y_i \geq t\right) \leq \underbrace{N\mathbb{P}(Y_1 \geq \eta t N)}_{(*)} + (\eta t N)^{-p} \cdot \underbrace{N \cdot \mathbb{E}[Y_1^p \cdot \mathbb{1}_{\{0 \leq Y_1 \leq \eta t N\}}]}_{(\times)} +$$

$$\underbrace{\exp\left(-\frac{(tN)^2}{c_p N \mathbb{E}[Y_1^2]}\right)}_{(\circ)}.$$

The first term $(*)$ can be bounded by Markov's inequality as

$$N\mathbb{P}\left(Y_1 \geq \eta t N\right) \leq N\mathbb{E}\left[|Y_1|^p\right] \cdot \eta^{-p} \cdot (t \cdot N)^{-p} = C_p \cdot t^{-p} \cdot N^{-p+1}.$$

Since $(\times)$ can obviously be bounded by m_p, the second term can be similarly bounded as

$$(\eta t N)^{-p} \cdot N\mathbb{E}\left[Y_1^p \cdot \mathbb{1}_{\{0 \leq Y_1 \leq \eta t N\}}\right] \leq (\eta t N)^{-p} \cdot N\mathbb{E}[|Y_1|^p] = C_p \cdot t^{-p} \cdot N^{-p+1}.$$

For the third term, we use the straightforward inequality for $a > 0$, $b > 0$ and $r > 0$,

$$\forall x \geq \max\left(\frac{|\ln(b)|}{r}, \frac{4r^2}{a^2}\right): \quad \exp(-ax) \leq bx^{-r}$$

and obtain with $r = p - 1$, $a = t^2/(c_p m_2)$ and $b = C_p t^{-p}$

$$\exp\left(-\frac{t^2}{c_p m_2}N\right) \leq C_p \cdot t^{-p} \cdot N^{-p+1}$$

for $N \geq \bar{N}(t)$ with $\bar{N}(t) := \max\left(\frac{|\ln(C_p) - p\ln(t)|}{p-1}, \frac{4(p-1)^2(c_p m_2)^2}{t^4}\right)$. $\square$

To obtain later on also a rate of convergence we extend inequality (2.2) and consider thus additionally some rate factor $N^{-\beta}$ in the decay:

Corollary 2.2.3.
Let $p \geq 2$ and $\{Y_N\}_{N \in \mathbb{N}}$ be a sequence of independent and identically distributed random variables with $\mathbb{E}[Y_1] = 0$, $m_p = \mathbb{E}[|Y_1|^p] < \infty$ and let $0 \leq \beta < \frac{1}{2}$. Then, for each $t > 0$, there exists some $\bar{N}(t)$ such that:

$$\forall N \geq \bar{N}(t): \quad \mathbb{P}\left(\frac{1}{N}\sum_{i=1}^{N} Y_i \geq tN^{-\beta}\right) \leq 3 \cdot C_p \cdot t^{-p} N^{-p(1-\beta)+1}. \qquad (2.3)$$

Again, $\bar{N}(t)$ is explicitly given in terms of the parameters.

Proof.
We again apply the Fuk-Nagaev inequality (2.1) and obtain for each N and given $t > 0$:

$$\mathbb{P}\left(\frac{1}{N}\sum_{i=1}^{N} \geq tN^{-\beta}\right) \leq \underbrace{N\mathbb{P}\left(Y_1 \geq \eta t N^{1-\beta}\right)}_{(*)} + (\eta t N^{1-\beta})^{-p}$$

$$\cdot N \cdot \underbrace{\mathbb{E}[Y_1^p \cdot \mathbb{1}_{\{0 \leq Y_1 \leq \eta t N^{1-\beta}\}}]}_{(\times)} + \underbrace{\exp\left(-\frac{t^2 N^{2-2\beta}}{c_p N \mathbb{E}[Y_1^2]}\right)}_{(\circ)}.$$

As before, the first term $(*)$ can be bounded by Markov's inequality as

$$N\mathbb{P}\left(Y_1 \geq \eta t N^{1-\beta}\right) \leq C_p \cdot t^{-p} \cdot N^{-p(1-\beta)+1}.$$

With $(\times)$ we proceed similar as in the previous proof and obtain the bound

$$(\eta t N^{1-\beta})^{-p} \cdot N\mathbb{E}[Y_1^p \cdot \mathbb{1}_{\{0 \leq Y_1 \leq \eta t N^{1-\beta}\}}] \leq (\eta t N^{1-\beta})^{-p} \cdot N\mathbb{E}[|Y_1|^p]$$
$$= C_p \cdot t^{-p} \cdot N^{-p(1-\beta)+1}.$$

For the third term, we now use a slightly more involved inequality for $a > 0, b > 0,$ $r > 0$ and $s > 0$:

$$\forall x \geq \max\left(\left[\frac{s}{2r}|\ln(b)|\right]^{2/s}, \left[\frac{4r}{as}\right]^{2/s}\right): \quad \exp(-ax^s) \leq bx^{-r},$$

and obtain with $r = p(1-\beta)+1$, $s = 2-2\beta$, $a = t^2/(c_p m_2)$ and $b = C_p t^{-p}$

$$\exp\left(-\frac{t^2}{c_p m_2}N^{2-2\beta}\right) \leq C_p \cdot t^{-p} \cdot N^{-p(1-\beta)+1}$$

and $\bar{N}(t)$ accordingly. $\qquad\square$

Remark 2.2.4.
Let us emphasize, as an immediate consequence of Corollary 2.2.3, it exists for each $t > 0$ *some* $\bar{N}(t)$ *such that*

$$\forall N \geq \bar{N}(t): \quad \mathbb{P}\left(\left|\frac{1}{N}\sum_{i=1}^{N} Y_i\right| \geq tN^{-\beta}\right) \leq 6 \cdot C_p \cdot t^{-p} \cdot N^{-p(1-\beta)+1}$$

holds.

2.3   Hoeffding Inequality

If the underlying sequence of r.v. $\{Y_N\}_{N \in \mathbb{N}}$ is also almost surely bounded, i.e. $a_i \leq Y_i \leq b_i$, $i = 1, \ldots, N$, we can apply the general Hoeffding inequality and obtain in contrast to the previous polynomial an exponential decay, i.e.:

Theorem 2.3.1. *(Hoeffding (1963), Theorem 2)*
Let $Y_1, \ldots, Y_N$ be independent random variables such that $a_i \leq Y_i \leq b_i$, $i = 1, \ldots, N$ holds. Then,

$$\forall t > 0: \quad \mathbb{P}\left(\frac{1}{N}\sum_{i=1}^{N} Y_i - \mathbb{E}\left[\frac{1}{N}\sum_{i=1}^{N} Y_i\right] \geq t\right) \leq \exp\left(-\frac{2N^2 t^2}{\sum_{i=1}^{N}(b_i - a_i)^2}\right).$$

Later on, an immediate consequence of Theorem 2.3.1 for $a_i = 0$, $b_i = 1$ is needed. Hence, we provide here an explicit version:

Corollary 2.3.2.
Let $Y_1, \ldots, Y_N$ be independent random variables such that $0 \leq Y_i \leq 1$, $i = 1, \ldots, N$ holds. Then,

$$\forall t > 0: \quad \mathbb{P}\left(\frac{1}{N}\sum_{i=1}^{N} Y_i - \mathbb{E}\left[\frac{1}{N}\sum_{i=1}^{N} Y_i\right] \geq t\right) \leq \exp\left(-2Nt^2\right).$$

Remark 2.3.3.
Let us emphasize that as an immediate consequence

$$\forall t > 0: \quad \mathbb{P}\left(\left|\frac{1}{N}\sum_{i=1}^{N} Y_i - \mathbb{E}\left[\frac{1}{N}\sum_{i=1}^{N} Y_i\right]\right| \geq t\right) \leq 2\exp\left(-2Nt^2\right)$$

holds straightforwardly.

2.4   Central Limit Theorem and Berry-Esseen Theorem

In order to obtain non parametric noise considering confidence intervals in Chapter 5 we apply, first, the well-known CLT approximation for the empirical variance $\hat{\sigma}_i^2$.

Theorem 2.4.1.
Let $\{Y_N\}_{N\in\mathbb{N}}$ *be a sequence of independent and identically distributed random variables with* $\mu = \mathbb{E}[Y_1] \in \mathbb{R}$ *and* $\sigma^2 = Var(Y_1) \in (0, \infty)$. *Then, for all* $z \in \mathbb{R}$,

$$\lim_{N\to\infty} \mathbb{P}\left(\frac{\sum_{i=1}^N Y_i - N\cdot\mu}{\sqrt{N}\cdot\hat{\sigma}} \leq z\right) = \Phi(z)$$

holds, for $\hat{\sigma}^2 := \frac{1}{N}\sum_{i=1}^N \left(Y_i - \frac{1}{N}\sum_{i=1}^N Y_i\right)^2$.

Proof.
By the standard CLT (see e.g. Klenke (2014), Theorem 15.37) we immediately obtain

$$\sqrt{N}\left(\sum_{i=1}^N Y_i - \mu\right) \xrightarrow[N\to\infty]{\mathcal{D}} \mathcal{N}(0, \sigma^2),$$

whereby $\mathcal{D}$ denotes the convergence in distribution. According to the weak law of large numbers $\hat{\sigma}^2 \xrightarrow[N\to\infty]{\mathbb{P}} \sigma^2$ holds ($\mathbb{P}$ denotes the convergence in probability) and with the continuous mapping Theorem $\hat{\sigma} \xrightarrow[N\to\infty]{\mathbb{P}} \sigma$. Finally, Slutsky's Theorem yields

$$\frac{\sqrt{N}\left(\sum_{i=1}^N Y_i - \mu\right)}{\hat{\sigma}} \xrightarrow[N\to\infty]{\mathcal{D}} \mathcal{N}(0, 1)$$

and thus the claim is proven. $\square$

Now, to obtain also rates of convergence the well-known Berry-Esseen Theorem is of interest since it specifies explicitly the speed of convergence towards the standard normal distribution in the CLT. According to Berry (1941) as well as Esseen (1942) the Berry-Essen Theorem states:

Theorem 2.4.2. *(Klenke (2014), Theorem 15.51)*
Let $\{Y_N\}_{N\in\mathbb{N}}$ *be a sequence of independent and identically distributed random variables with* $\mathbb{E}[Y_1] = 0$, $\sigma^2 = \mathbb{E}[Y_1^2] \in (0, \infty)$ *and* $\kappa^3 := \mathbb{E}[|Y_1|^3] < \infty$. *Then,*

$$\forall N \in \mathbb{N}: \quad \sup_{x\in\mathbb{R}} \left|\mathbb{P}\left(\frac{\sum_{i=1}^N Y_i}{\sqrt{N\sigma^2}} \leq x\right) - \Phi(x)\right| \leq \frac{C\cdot\kappa^3}{\sigma^3\sqrt{N}}$$

holds, for some constant $C \in \mathbb{R}$ and Φ denotes the cdf of the standard normal distribution.

Originally, the Berry-Essen constant C is given by 7.59 (cf. Esseen (1942)). Esseen (1956) himself determined an asymptotically best lower bound for C, i.e. $C \geq (3 + \sqrt{10})/6\sqrt{2\pi} \approx 0.410$. Note, that an optimal estimate for C is still an active field of research.

Almost Sure Convergence of Moment-Based Estimators

3

In the following, we investigate the moment-based problem, cf. (1.10), especially the so far unexamined almost sure convergence. The whole chapter is based on the results of Klein and Werner (2023b). The corresponding research question is rather easy to postulate, i.e.: Which minimal growth condition on N_1 and N_2 leads to an uniform almost sure convergence of $\bar{\gamma}_{N_1, N_2}$ towards γ? Furthermore, the speed of convergence will also be examined. As already mentioned, cf. Table 1.1, these are to the best of our knowledge so far unanswered questions.

This chapter is subdivided as follows: In Section 3.1 we discuss a variety of assumptions which are necessary to derive the main results. In Section 3.2 we will provide examples when these assumptions are naturally satisfied and how the asymptotic bias behaves in such settings. Several function classes like polynomials will be examined. In particular, the commonly known and already addressed probability of a large loss problem will be examined exhaustively since it describes according to Hong and Juneja (2009) and Gordy and Juneja (2010) a major risk measure in this moment-based setting. The main novel results start in Section 3.3, where we provide a significant improvement of Rainforth et al. (2018), Theorem 2, under similarly weak assumptions. Afterwards in Section 3.4, we provide the two main theorems, Theorem 3.4.1 and Theorem 3.4.4, which state the convergence in probability and the almost sure convergence of $\bar{\gamma}_{N_1, N_2}$. While Theorem 3.4.1 focuses on the strong consistency of the estimator, Theorem 3.4.4 additionally provides rates for both types of convergence. These convergence rates crucially depend on the asymptotic behaviour of the bias and more or less recover all known rates for other modes of convergence. Our numerical investigations on these theoretical complexities are summarized in Chapter 6, in particular in Section 6.1.

M. Klein, *Nested Simulations: Theory and Application*, Mathematische Optimierung und Wirtschaftsmathematik | Mathematical Optimization and Economathematics, https://doi.org/10.1007/978-3-658-43853-1_3

3.1   Assumptions

The main quantities in our analysis are the random variables X, ϵ_{N_2}, $G(X)$, and $G(X + \epsilon_{N_2})$. Our very weak basic assumption on these quantities is that they are well-defined, i.e. we make the assumption

$$V \in L^1(\Omega, \mathcal{F}, \mathbb{P}), \quad G(X) \in L^1(\Omega, \mathcal{F}, \mathbb{P}), \quad \forall N_2 : G(X + \epsilon_{N_2}) \in L^1(\Omega, \mathcal{F}, \mathbb{P}), \tag{A}$$

whereby the first assumption especially implies $X \in L^1(\Omega, \mathcal{F}, \mathbb{P})$ and $\epsilon_{N_2} \in L^1(\Omega, \mathcal{F}, \mathbb{P})$ for each N_2, see Proposition 1.4.1. We also note that due to Proposition 1.4.1, under Assumption (A), the law of large numbers yields

$$\epsilon_{N_2} \xrightarrow[N_2 \to \infty]{\text{a.s.}} 0 \quad \text{and} \quad \epsilon_{N_2} \xrightarrow[N_2 \to \infty]{L^1} 0.$$

Note, that in our setup no assumptions concerning the conditioning variable Z are formulated.

Continuity and Integrability Assumptions

Since we cannot expect any convergence result, if X has mass at discontinuity points of G, we need to require the following weak regularity of G, i.e. we assume

$$\mathbb{P}(X \in D_G) = 1, \quad \text{where } D_G := \{x \in \mathbb{R} \mid G \text{ is continuous at } x\}. \tag{A0}$$

We note that under Assumptions (A) and (A0), we have $G(X + \epsilon_{N_2}) \xrightarrow[N \to \infty]{\text{a.s.}} G(X)$, as $X + \epsilon_{N_2} \xrightarrow[N \to \infty]{\text{a.s.}} X$ due to Proposition 1.4.1 and the continuous mapping theorem. Still, Assumptions (A) and (A0) are not sufficient for our purposes. Further assumptions on the regularity of G, especially on the moments of the involved quantities, i.e. on moments of $G(X + \epsilon_{N_2})$ and $G(X)$, are required. We therefore state the following two integrability assumptions for some $p \geq 1$:

$$\exists \bar{N}_2 \in \mathbb{N} : \sup_{N_2 \geq \bar{N}_2} \|G(X + \epsilon_{N_2})\|_{L^p} < \infty, \tag{A1,p}$$

and

$$\|G(X)\|_{L^p} < \infty. \tag{A2,p}$$

In the subsequent Proposition 3.1.1, we clarify the close relationship of the rather weak Assumption (A1,p) and Assumption (A2,p) to the seemingly stronger assumptions of *convergence in p-norm*, i.e.

$$\|G(X + \epsilon_{N_2}) - G(X)\|_{L^p} \xrightarrow[N_2 \to \infty]{} 0, \qquad \text{(CiN,}p)$$

as well as to the slightly weaker assumption of *convergence of p-norms*, i.e.

$$\|G(X + \epsilon_{N_2})\|_{L^p} \xrightarrow[N_2 \to \infty]{} \|G(X)\|_{L^p}. \qquad \text{(CoN,}p)$$

We emphasize that the relations given in Proposition 3.1.1 hold as we already have $G(X + \epsilon_{N_2}) \xrightarrow[N \to \infty]{\text{a.s.}} G(X)$ in our setup under Assumptions (A) and (A0).

Proposition 3.1.1.
Let Assumptions (A) and (A0) be satisfied. Then, the following relations hold:

1. *For all $p \geq 1$: (A2,p) and (CoN,p) hold $\Longleftrightarrow$ (A1,p) and (CiN,p) hold.*
2. *For all $p \geq 1$: (A2,p) and (CiN,p) hold $\Longrightarrow$ (A1,p) holds.*
3. *For all $p \geq 1$: (A1,p) holds $\Longrightarrow$ (A2,p) holds.*
4. *For all $p > 1$: (A1,p) holds $\Longrightarrow$ (CiN,p') holds for all $1 \leq p' < p$.*

Proof.
"1., $\Longrightarrow$": We note that (A2,p) and (CoN,p) imply (A1,p). Further, (CiN,p) follows from (CoN,p) by the Riesz convergence theorem, cf. Novinger (1972).
"1., $\Longleftarrow$": (A2,p) follows from (A1,p) together with (CiN,p) by the triangle inequality. The statement that (CiN,p) implies (CoN,p) is obvious.
"2.": This statement follows from 1. since (CiN,p) implies (CoN,p).
"3. and 4.": These statements follow from Elstrodt (2018) Exercise VI.5.1. $\qquad \square$

We observe that, basically, it is sufficient to work with Assumption (A1,p) for some $p > 1$ in the given context. Let us further remark that the assumptions considered in Proposition 3.1.1 are satisfied for large function classes like Lipschitz functions, bounded functions or polynomial bounded functions, see Proposition 3.2.3, Proposition 3.2.1 and Proposition 3.2.2 in Section 3.2 for more details. These classes especially cover the important examples of the large loss probability and the lower partial moment of first order.

Assumptions on the Bias

Instead of making assumptions on moments of $G(X + \epsilon_{N_2})$ and $G(X)$, we can alternatively assume that the *bias* $B_{N_2} := \mathbb{E}[G(X + \epsilon_{N_2})] - \mathbb{E}[G(X)]$ vanishes asymptotically, similar to Andradóttir and Glynn (2016) or Rainforth et al. (2016, 2018):

$$\mathbb{E}[G(X + \epsilon_{N_2})] \xrightarrow[N_2 \to \infty]{} \mathbb{E}[G(X)]. \tag{VB}$$

In Rainforth et al. (2016, 2018), the authors assume that *expected continuity* holds, i.e. it is assumed that

$$||G(X + \epsilon_{N_2}) - G(X)||_{L^1} \xrightarrow[N_2 \to \infty]{} 0, \tag{EC}$$

which in our notation coincides with (CiN,1); an assumption which immediately implies (VB). In Rainforth et al. (2018), (CiN,1) is considered to be a kind of minimal assumption and used in the derivation of Rainforth et al. (2018), Theorem 2, which we discuss in Section 3.3 in more detail.

Finally, similar to Andradóttir and Glynn (2016), to obtain asymptotic convergence rates, we need the stronger assumption

$$N_2^{\beta} \cdot |\mathbb{E}[G(X + \epsilon_{N_2})] - \mathbb{E}[G(X)]| \xrightarrow[N_2 \to \infty]{} 0 \tag{VB,β}$$

for some $\beta \geq 0$. We note that the original assumption (VB) coincides with (VB,β) for $\beta = 0$. More details on the order of the bias for a variety of function classes is given in Section 3.2.

3.2   Bias Considerations

In the following paragraphs, we will consider specific function classes in combination with moment assumptions for V and investigate, which of the assumptions considered in Section 3.1 are satisfied for these combinations. This question is interesting in its own and is for example raised in Rainforth et al. (2016), ahead of Theorem 1 (cf. p. 3), especially concerning expected continuity. In summary, we provide a few new results on the bias for function classes not considered before (like bounded functions or polynomials) which hold under weak assumptions. Most importantly, we show that polynomials enjoy a bias order of -1 under quite weak assumptions.

Function Classes Satisfying (VB)

Let us start with the following easy observation: In case G is bounded, no moment condition on V beyond the existence of the first moment is necessary to guarantee that the bias is vanishing asymptotically. However, in this case, no rate for the bias can be obtained.

Proposition 3.2.1.
Let Assumptions (A) *and* (A0) *be satisfied and let G be bounded. Then, for all $p \geq 1$ Assumptions* (A1,p), (A2,p), (CiN,p) *and* (CoN,p) *hold as well as* (VB) *is satisfied.*

Proof.
Assumptions (A1,p) and (A2,p) obviously hold for all p due to the boundedness of G. The rest follows by Proposition 3.1.1(3). $\square$

We note that the large loss probability is contained within this class of bounded functions.

Similarly easy, we can also cover the situation that G grows at most polynomially:

Proposition 3.2.2.
Let Assumptions (A) *and* (A0) *be satisfied and let G be bounded as follows: $\forall x \in \mathbb{R}: |G(x)| \leq |x|^k + c$ for some $k, c \geq 0$. Then $V \in L^{pk} (\Omega, \mathcal{F}, \mathbb{P})$ implies* (A1,p) *and* (A2,p)*, and* (CoN,p') *and* (CiN,p') *hold for all $1 \leq p' < p$. Further,* (VB) *holds.*

Proof.
We have

$$\mathbb{E}[|G(X)|^p] \leq \mathbb{E}[(|X|^k + c)^p] = |||X|^k + c||_{L^p}^p \leq (||X^k||_{L^p} + c)^p$$
$$\leq 2^p(||X^k||_{L^p}^p + c^p) = 2^p(\mathbb{E}[|X|^{pk}] + c^p),$$

which obviously implies (A2,p). Replacing X by $X + \epsilon_{N_2}$ yields (A1,p), the rest follows again from Proposition 3.1.1(3).

This class contains for instance the lower partial moment, where $k = 1$ can be chosen. Let us further remark that this situation is e.g. given in Hong and Juneja (2009) for $k = 3$ and $p = 4/3$, where G is assumed to be piecewise thrice continuously differentiable with bounded third derivative. However, as we do not exploit differentiability of G here, no rate for the vanishing bias can be obtained in this setup.

Function Classes Satisfying (VB,β) for $0 \leq \beta < 1/2$

For the large class of globally Lipschitz continuous functions, which for example also contain the lower partial moment, we obtain the following result. This result significantly improves upon the result from the previous subsection, as now an order of (almost) $-1/2$ can be obtained for the bias.

Proposition 3.2.3.
Let Assumptions (A) and (A0) be satisfied and let G be Lipschitz continuous with constant $L > 0$. Then $V \in L^p (\Omega, \mathcal{F}, \mathbb{P})$ for some $p \geq 1$ implies (A1,p), (A2,p), (CiN,p), and (CoN,p) and thus (VB). For $p \geq 2$, $V \in L^p (\Omega, \mathcal{F}, \mathbb{P})$ additionally implies (VB,β) for all $0 \leq \beta < 1/2$.

Proof.
We start by noting that

$$||G(X)||_{L^p} \leq ||G(X) - G(0)||_{L^p} + ||G(0)||_{L^p} \leq L||X||_{L^p} + |G(0)|$$

and

$$||G(X + \epsilon_{N_2})||_{L^p} \leq ||G(X + \epsilon_{N_2}) - G(X)||_{L^p} + ||G(X)||_{L^p} \leq L||\epsilon_{N_2}||_{L^p} + ||G(X)||_{L^p}.$$

Hence, (A1,p) and (A2,p) follow immediately by the moment calculations in Proposition 1.4.1. Since

$$||G(X + \epsilon_{N_2}) - G(X)||_{L^p} \leq L||\epsilon_{N_2}||_{L^p},$$

the remaining claims for $p > 1$ follow by the conditional Rosenthal inequalities in Proposition 1.4.1. For $p = 1$, the remaining claims follow due the strong law of large numbers given in Proposition 1.4.1.   □

While the result as such is not new, it especially helps to put assumptions into context. Therefore, we have included it in this list of results.

Function Classes Satisfying (VB,β) for $0 \leq \beta < 1$

If a better order of convergence for the bias is sought for (more precisely an order of (almost) -1), both G and V have to satisfy further assumptions. Our main contribution here is to show that, given sufficient moments of V, polynomials allow a bias of order (almost) -1, see Proposition 3.2.2 for more details. For this purpose, we first recall a related result which has appeared in the literature several times in slightly different variations, see Rainforth et al. (2018), Theorem 3, Liu et al. (2022), Lemma 1, or Hong and Juneja (2009), Section 2.

Proposition 3.2.4.
Let Assumptions (A) *and* (A0) *be satisfied, let* $V \in L^2(\Omega, \mathcal{F}, \mathbb{P})$ *and let* G *be twice continuously differentiable with* $\sup_{x \in \mathbb{R}} |G''(x)| \leq M < \infty$. *Then, it holds*

$$|\mathbb{E}[G(X + \epsilon_{N_2})] - \mathbb{E}[G(X)]| \leq \frac{1}{2} M \cdot \|V\|_{L^2}^2 \cdot N_2^{-1},$$

i.e. (VB,β) *holds for all* $0 \leq \beta < 1$.

Similarly, if Assumptions (A) *and* (A0) *are satisfied,* $V \in L^3(\Omega, \mathcal{F}, \mathbb{P})$, G *is three times continuously differentiable with* $\sup_{x \in \mathbb{R}} |G'''(x)| \leq M < \infty$ *and* $\mathbb{E}[|G''(X)|V^2] < \infty$ *is satisfied, then* (VB,β) *holds for all* $0 \leq \beta < 1$.

Proof.
Using Taylor's Theorem for some ξ between X and $X + \epsilon_{N_2}$ yields

$$\mathbb{E}[G(X + \epsilon_{N_2})] - \mathbb{E}[G(X)] = \mathbb{E}\left[G'(X) \cdot \epsilon_{N_2} + \frac{1}{2} \cdot G''(\xi) \cdot \epsilon_{N_2}^2\right].$$

Further,

$$\mathbb{E}\left[G'(X) \cdot \epsilon_{N_2}\right] = \mathbb{E}\left[\mathbb{E}\left[G'(X) \cdot \epsilon_{N_2} \mid Z\right]\right] = \mathbb{E}\left[G'(X) \cdot \mathbb{E}\left[\epsilon_{N_2} \mid Z\right]\right] = 0,$$

and

$$|\mathbb{E}\left[G''(\xi) \cdot \epsilon_{N_2}^2\right]| \leq \mathbb{E}\left[|G''(\xi) \cdot \epsilon_{N_2}^2|\right] \leq M \cdot \mathbb{E}\left[\epsilon_{N_2}^2\right].$$

With Proposition 1.4.1, the first claim follows. The second claim follows similarly with a Taylor approximation of third order, noting that

$$|\mathbb{E}[G''(X) \cdot \epsilon_{N_2}^2]| = |\mathbb{E}[G''(X) \cdot \mathbb{E}[\epsilon_{N_2}^2 \mid Z]]| \leq \frac{1}{N_2} \mathbb{E}[|G''(X)| \cdot \mathbb{E}[V^2 \mid Z]] = \frac{1}{N_2} \mathbb{E}[|G''(X)|V^2].$$

$\square$

In the following proposition, we show that the order of the bias is (almost) -1, if G is a polynomial. This constitutes a new result and significantly enlarges the class of functions G allowing for a fast decay of the bias. We emphasize that for polynomials less restrictive assumptions concerning V have to be satisfied, compared to the case of the large loss probability or the lower partial moment.

Proposition 3.2.5.
Let Assumptions (A) *and* (A0) *be satisfied, then it holds:*

1. *If* $V \in L^1(\Omega, \mathcal{F}, \mathbb{P})$ *and* G *is an affine function, then* $|\mathbb{E}[G(X + \epsilon_{N_2})] - \mathbb{E}[G(X)]| = 0$.
2. *If* $V \in L^p(\Omega, \mathcal{F}, \mathbb{P})$ *and* G *is a polynomial of degree* p *with* $p \geq 2$, *then, for some* $C_{G,p} \geq 0$, *it holds that*

$$|\mathbb{E}[G(X + \epsilon_{N_2})] - \mathbb{E}[G(X)]| \leq \frac{1}{2} \cdot C_{G,p} \cdot \|V\|_{L^2}^2 \cdot N_2^{-1},$$

i.e. (VB,β) *holds for all* $0 \leq \beta < 1$.

Proof.
The first statement is obvious. For $p = 2$, the second statement follows from Proposition 3.2.4. For $p \geq 3$ it is sufficient to consider the monomial case $G(x) = x^p$, the general claim then follows straightforwardly. In this case, we consider the equality

$$|\mathbb{E}[G(X + \epsilon_{N_2})] - \mathbb{E}[G(X)]| = |\mathbb{E}[(X + \epsilon_{N_2})^p - X^p]| = |\mathbb{E}[\sum_{k=0}^{p-1} a_k X^k \epsilon_{N_2}^{p-k}]|$$

with $a_k = \binom{p}{k}$, where we emphasize that the sum only runs to $p - 1$. Similarly to the proof of Proposition 3.2.4 we have for $k = p - 1$ that

$$\mathbb{E}[X^{p-1}\epsilon_{N_2}] = \mathbb{E}[X^{p-1}\mathbb{E}[\epsilon_{N_2} \mid Z]] = 0,$$

i.e. the term for $k = p - 1$ in $\mathbb{E}[\sum_{k=0}^{p-1} a_k X^k \epsilon_{N_2}^{p-k}]$ vanishes. For the term for $k = p - 2$ in this sum, we note that it further holds for $k = p - 2$ that

$$\mathbb{E}[X^{p-2}\epsilon_{N_2}^2] = \mathbb{E}[X^{p-2}\mathbb{E}[\epsilon_{N_2}^2 \mid Z]] = \frac{1}{N_2} \cdot \mathbb{E}[X^{p-2}(\mathbb{E}[V^2 \mid Z] - X^2)]$$

according to Proposition 1.4.1. By definition of $X = \mathbb{E}[V \mid Z]$ we have that

$$\mathbb{E}[X^{p-2}\mathbb{E}[V^2 \mid Z]] = \mathbb{E}[\mathbb{E}[V \mid Z]^{p-2}\mathbb{E}[V^2 \mid Z]].$$

We now use the Hölder inequality $\mathbb{E}[|RS|] \le \mathbb{E}[|R|^r]^{1/r} \cdot \mathbb{E}[|S|^s]^{1/s}$ for $\frac{1}{r} + \frac{1}{s} = 1$, where

$$R = \mathbb{E}[V \mid Z]^{p-2}, \quad S = \mathbb{E}[V^2 \mid Z], \quad r = p/(p-2), \quad \text{and } s = p/2.$$

Since $\mathbb{E}[|R|^r]^{1/r} = \mathbb{E}[|\mathbb{E}[V \mid Z]|^p]^{(p-2)/p} \le ||X||_{L^p}^{p-2}$ as well as $\mathbb{E}[|S|^s]^{1/s} = \mathbb{E}[|\mathbb{E}[V^2 \mid Z]|^{p/2}]^{2/p} \le ||X||_{L^p}^2$ (both inequalities are due to the conditional Jensen inequality) we obtain

$$|\mathbb{E}[X^{p-2}\epsilon_{N_2}^2]| \le \mathbb{E}[|X^{p-2}\epsilon_{N_2}^2|] \le \frac{1}{N_2} \cdot \left(\mathbb{E}[|X^{p-2}\mathbb{E}[V^2 \mid Z]|] + ||X||_{L^p}^p\right) \le \frac{2}{N_2}||X||_{L^p}^p$$

and thus an upper bound on the term for $k = p - 2$ in $\mathbb{E}[\sum_{k=0}^{p-2} a_k X^k \epsilon_{N_2}^{p-k}]$.

For the remaining terms in $\mathbb{E}[\sum_{k=0}^{p-3} a_k X^k \epsilon_{N_2}^{p-k}]$ we proceed similarly and obtain that these terms vanish faster than $1/N_2$; we only briefly sketch the case for $k = p - 3$, all other terms can be treated in the same way:

$$\mathbb{E}[|X^{p-3}\epsilon_{N_2}^3|] \le \mathbb{E}[|X^{p-3}|\mathbb{E}[|\epsilon_{N_2}|^3 \mid Z]]$$
$$\le C_p \cdot \mathbb{E}[|X^{p-3}|\left(N_2^{-2}\mathbb{E}[|W_1|^3 \mid Z] + N_2^{-3/2}\mathbb{E}[|W_1|^2 \mid Z]^{3/2}\right)].$$

Here, the second inequality is due to the Rosenthal inequality from Proposition 1.4.1. Both terms can now be upper bounded as above by the Hölder inequality which closes the proof. $\qquad\square$

For completeness, let us recall that Liu et al. (2022), Lemma 1, covers five more function classes, all with an order of -1 for the bias – however, under much stronger assumptions than assumed here.

Bias Considerations for the Large Loss Probability

In the following, we provide some results on the asymptotic bias in case of $G(X) = \mathbb{1}_{\{X>c\}}$ for a given $c \in \mathbb{R}$. We especially distinguish the continuous case from the discrete case, similar to Lee (1998), Andradóttir and Glynn (2016) and Glynn and Lee (2003).

We start with a very general result which yields an asymptotic order of (almost) $-1/3$, if second moments exist and which improves to (almost) $-1/2$ if moments of very high order exist. This result is applicable in almost all situations, as any cumulative distribution function is differentiable almost everywhere.

Proposition 3.2.6.
Let $V \in L^p(\Omega, \mathcal{F}, \mathbb{P})$, let $G(X) = \mathbb{1}_{\{X>c\}}$ for some $c \in \mathbb{R}$ and let F_X be differentiable at c. For $1 < p \leq 2$, Assumption (VB,β) is satisfied for all $0 \leq \beta < \frac{p-1}{p+1}$. For $p > 2$, Assumption (VB,β) is satisfied for all $0 \leq \beta < \frac{p}{2(p+1)}$.

Proof.
Let us assume that $F_X'(c) > 0$ holds and let us consider

$$\mathbb{E}\left[\mathbb{1}_{\{X+\epsilon_{N_2}>c\}}\right] - \mathbb{E}\left[\mathbb{1}_{\{X>c\}}\right] = \mathbb{P}\left(X + \epsilon_{N_2} > c\right) - \mathbb{P}\left(X > c\right) = F_X(c) - F_{X+\epsilon_{N_2}}(c).$$

Using the well-known relationship between a distorted distribution function $F_{X+\epsilon_{N_2}}$ and a true distribution function F_X (see for instance Lemma 3 in Petrov (1975)), we obtain for all N_2:

$$F_X(c) - F_{X+\epsilon_{N_2}}(c) \leq F_X(c) - \left(F_X(c - N_2^{-\beta}) - \mathbb{P}\left(|\epsilon_{N_2}| \geq N_2^{-\beta}\right)\right).$$

Since F_X is assumed to be differentiable at c, it further holds for $s \geq 0$:

$$F_X(c - s) = F_X(c) - s \cdot F_X'(c) - s \cdot R(s),$$

with residual term $R(s) \to 0$ for $s \to 0$. Hence, for N_2 large enough, we have $|R(N_2)| \leq F_X'(c)$ and thus, we obtain for $N_2 \geq \bar{N}_2$ with sufficiently large $\bar{N}_2$:

$$F_X(c) - \left(F_X(c - N_2^{-\beta}) - \mathbb{P}\left(|\epsilon_{N_2}| \geq N_2^{-\beta}\right)\right) \leq 2 \cdot F_X'(c) \cdot N_2^{-\beta} + \mathbb{P}\left(|\epsilon_{N_2}| \geq N_2^{-\beta}\right).$$

Since $\epsilon_{N_2} = \frac{1}{N_2} \sum_{j=1}^{N_2} W_j$ is the sum of conditionally independent random variables W_j, we need to consider the conditional probability $\mathbb{P}\left(|\epsilon_{N_2}| \geq N_2^{-\beta} \mid Z\right)$ for corresponding upper bounds, i.e. we use

$$\mathbb{P}\left(|\epsilon_{N_2}| \geq N_2^{-\beta}\right) = \mathbb{E}\left[\mathbb{P}\left(|\epsilon_{N_2}| \geq N_2^{-\beta} \mid Z\right)\right]$$

and apply the conditional Markov inequality to get

$$\mathbb{P}\left(|\epsilon_{N_2}| \geq N_2^{-\beta}\right) = \mathbb{E}\left[\mathbb{P}\left(|\epsilon_{N_2}| \geq N_2^{-\beta} \mid Z\right)\right] \leq \mathbb{E}[\mathbb{E}[|\epsilon_{N_2}|^p \mid Z] \cdot N_2^{\beta p}].$$

If, on the one hand, $1 < p \leq 2$ holds we obtain by Proposition 1.4.1

$$\mathbb{E}[|\epsilon_{N_2}|^p \mid Z] \cdot N_2^{\beta p} \leq C_p \cdot \mathbb{E}[|W_j|^p \mid Z] \cdot N_2^{-p(1-\beta)+1},$$

with $W_j = V_j - X$ and C_p depending on p only. In summary, we get

$$F_X(c) - \left(F_X(c - N_2^{-\beta}) - \mathbb{P}\left(|\epsilon_{N_2}| \geq N_2^{-\beta}\right)\right) \leq 2 \cdot F_X'(c) \cdot N_2^{-\beta} + C_p \cdot \mathbb{E}[|W_j|^p] \cdot N_2^{-p(1-\beta)+1}.$$

If $\beta < \frac{p-1}{p+1}$, then $C_p \cdot \mathbb{E}[|W_j|^p] \cdot N_2^{-p(1-\beta)+1} \leq 2 \cdot F_X'(c) \cdot N_2^{-\beta}$ holds for large enough N_2, hence

$$F_X(c) - \left(F_X(c - N_2^{-\beta}) - \mathbb{P}\left(|\epsilon_{N_2}| \geq N_2^{-\beta}\right)\right) \leq 4 \cdot F_X'(c) \cdot N_2^{-\beta}.$$

If, on the other hand, $p > 2$ holds, Proposition 1.4.1 yields

$$\mathbb{E}[|\epsilon_{N_2}|^p \mid Z] \cdot N_2^{\beta p} \leq C_p \cdot \left(N_2^{-p+1} \cdot \mathbb{E}[|W_j|^p \mid Z] + N_2^{-p/2} \cdot \mathbb{E}[W_j^2 \mid Z]\right) \cdot N_2^{\beta p}$$

$$\leq C_p \cdot N_2^{-(p/2)+\beta p} \cdot \left(N_2^{-p/2+1} \cdot \mathbb{E}[|W_j|^p \mid Z] + \mathbb{E}[W_j^2 \mid Z]\right).$$

In this case we obtain similarly

$$F_X(c) - \left(F_X(c - N_2^{-\beta}) - \mathbb{P}\left(|\epsilon_{N_2}| \geq N_2^{-\beta}\right)\right) \leq$$

$$\leq 2 \cdot F_X'(c) \cdot N_2^{-\beta} + C_p \cdot N_2^{-(p/2)+\beta p} \cdot \left(N_2^{-p/2+1} \cdot \mathbb{E}[|W_j|^p] + \mathbb{E}[W_j^2]\right).$$

Then, for $\beta < \frac{p}{2(p+1)}$ we have $C_p \cdot N_2^{-(p/2)+\beta p} \cdot \left(N_2^{-p/2+1} \cdot \mathbb{E}[|W_j|^p] + \mathbb{E}[W_j^2] \right) \leq$
$2 \cdot F_X'(c) \cdot N_2^{-\beta}$ for large enough N_2 and thus

$$F_X(c) - \left(F_X(c - N_2^{-\beta}) - \mathbb{P}\left(|\epsilon_{N_2}| \geq N_2^{-\beta} \right) \right) \leq 4 \cdot F_X'(c) \cdot N_2^{-\beta}.$$

Since the remaining inequality follows analogously, this shows the claim. The slight adjustment of the proof for the case $F_X'(c) = 0$ is obvious. $\qquad\square$

If we additionally assume that the p-th conditional moment of V is uniformly bounded, we obtain the following improved version which is based on a conditional version of the Fuk-Nagaev inequality instead of the conditional Rosenthal inequality. While for $p = 2$ the rate is not improved, we already obtain an asymptotic order of (almost) $-1/2$ for $p \geq 3$.

Proposition 3.2.7.
Let $V \in L^p(\Omega, \mathcal{F}, \mathbb{P})$, $p \geq 2$, $\| \mathbb{E}[|V|^p \mid Z] \|_{L^\infty} = M < \infty$ and let $G(X) = \mathbb{1}_{\{X > c\}}$ for some $c \in \mathbb{R}$. If F_X is differentiable at c, then Assumption (VB,β) is satisfied for all $0 \leq \beta < \min(\frac{1}{2}, \frac{p-1}{p+1})$.

Proof.
The first part of this proof is identical to proof of Proposition 3.2.6, up to where we have applied the conditional Rosenthal inequality to $\mathbb{P}\left(|\epsilon_{N_2}| \geq N_2^{-\beta} \mid Z \right)$. Since $\| \mathbb{E}[|V|^p \mid Z] \|_{L^\infty} = M < \infty$ holds, a similar bound holds for the second conditional moment due to the conditional Hölder inequality. If $\beta < 1/2$ we can apply Corollary 2.2.3 to $\mathbb{P}\left(|\epsilon_{N_2}| \geq N_2^{-\beta} \mid Z \right)$ and obtain correspondingly that there exists a random variable $\bar{N}_2(Z)$ such that almost surely

$$\forall N_2 \geq \bar{N}_2(Z): \quad \mathbb{P}\left(|\epsilon_{N_2}| \geq N_2^{-\beta} \mid Z \right) \leq 3 \cdot C_p^Z \cdot N_2^{-p(1-\beta)+1},$$

with $C_p^Z := \mathbb{E}[|W_1|^p|Z] \cdot \eta^{-p}$. To continue, we need that the random variable $\bar{N}_2(Z)$ can be chosen in such a way that it is bounded by some constant $\hat{N}_2 \in \mathbb{N}$ which follows directly from the specific choice of $\bar{N}_2(Z)$ in Corollary 2.2.3 (for $t = 1$) due to the assumption $\| \mathbb{E}[|V|^p \mid Z] \|_{L^\infty} = M < \infty$. We thus obtain an $\hat{N}_2 \in \mathbb{R}$ such that for all $N_2 \geq \hat{N}_2$:

$$F_X(c) - \left(F_X(c - N_2^{-\beta}) - \mathbb{P}\left(|\epsilon_{N_2}| \geq N_2^{-\beta}\right)\right) = 2F_X'(c)N_2^{-\beta} + \mathbb{E}\left[\mathbb{P}\left(|\epsilon_{N_2}| \geq N_2^{-\beta} \mid Z\right)\right]$$
$$\leq 2 \cdot F_X'(c) \cdot N_2^{-\beta} + 3 \cdot C_p \cdot N_2^{-p(1-\beta)+1}.$$

If $\beta < \frac{p-1}{p+1}$, then $3 \cdot C_p \cdot N_2^{-p(1-\beta)+1} \leq 2 \cdot F_X'(c) \cdot N_2^{-\beta}$ holds for large enough N_2, hence

$$F_X(c) - \left(F_X(c - N_2^{-\beta}) - \mathbb{P}\left(|\epsilon_{N_2}| \geq N_2^{-\beta}\right)\right) \leq 4 \cdot F_X'(c) \cdot N_2^{-\beta}.$$

The remainder of the proof runs analogously. $\qquad\square$

Let us point out that the improvement of Proposition 3.2.7 versus Proposition 3.2.6 relies on the fact that ϵ_{N_2} is an average of (conditionally) independent variables, which has an immediate impact on the tail behaviour of ϵ_{N_2}.

If we move away from continuous random variables to discrete ones, the above result can be strengthened. More exactly, we assume that F_X is (locally) flat around c; an assumption which is satisfied for almost all c in case that X follows a discrete (i.e. finite or countably infinite) distribution, or if the density of X vanishes in a neighborhood of c. In (almost) this setup, Glynn and Lee (2003) consider the case that the moment generating function of $V \mid Z$ exists and is uniformly bounded, and derive an order of $\exp(-aN_2)$ for some $a > 0$. Therefore, we here consider the more general case that only some conditional moment is uniformly bounded, and accordingly (only) obtain a polynomial decay of order of either $N_2^{-p/2}$ or N_2^{-p+1}.

Proposition 3.2.8.
Let $V \in L^p(\Omega, \mathcal{F}, \mathbb{P})$, $p > 1$ and let $G(X) = \mathbb{1}_{\{X > c\}}$ for some $c \in \mathbb{R}$. Let further F_X be locally flat at c, i.e. there exists a $\delta > 0$ such that $F_X(c - \delta) = F_X(c + \delta)$. If $1 < p \leq 2$, then Assumption (VB,β) is satisfied for all $0 \leq \beta < p - 1$. If $p > 2$, then Assumption (VB,β) is satisfied for all $0 \leq \beta < p/2$; if additionally $\|\mathbb{E}[|V|^p \mid Z]\|_{L^\infty} = M < \infty$ for some $M > 0$, then Assumption (VB,β) is satisfied for all $0 \leq \beta < p - 1$.

Proof.
Similar to the proof of Proposition 3.2.6 we consider

$$\mathbb{E}\left[\mathbb{1}_{\{X + \epsilon_{N_2} > c\}}\right] - \mathbb{E}\left[\mathbb{1}_{\{X > c\}}\right] = \mathbb{P}\left(X + \epsilon_{N_2} > c\right) - \mathbb{P}(X > c) = F_X(c) - F_{X + \epsilon_{N_2}}(c).$$

Using once again Lemma 3 in Petrov (1975), we obtain:

$$F_X(c) - F_{X+\epsilon_{N_2}}(c) \leq F_X(c) - \left(F_X\left(c - \frac{\delta}{2}\right) - \mathbb{P}\left(|\epsilon_{N_2}| \geq \frac{\delta}{2}\right)\right).$$

Due to local flatness at c, it further holds:

$$F_X(c) - \left(F_X\left(c - \frac{\delta}{2}\right) - \mathbb{P}\left(|\epsilon_{N_2}| \geq \frac{\delta}{2}\right)\right) = \mathbb{P}\left(|\epsilon_{N_2}| \geq \frac{\delta}{2}\right).$$

Similar to the proof of Proposition 3.2.7, from $\|\mathbb{E}[|V|^p \mid Z]\|_{L^\infty} = M < \infty$ we obtain by Corollary 2.2.2 an $\bar{N}_2$ such that for all $N_2 \geq \bar{N}_2$:

$$\mathbb{P}\left(|\epsilon_{N_2}| \geq \frac{\delta}{2}\right) = \mathbb{E}\left[\mathbb{P}\left(|\epsilon_{N_2}| \geq \frac{\delta}{2}\Big| Z\right)\right] \leq 3 \cdot C_p \cdot \left(\frac{\delta}{2}\right)^{-p} \cdot N_2^{-p+1}.$$

The discussion regarding the existence of $\bar{N}_2$ is analogous to the discussion in the proof of Proposition 3.2.7. This inequality yields the third claim.

Without the assumption of V having uniformly bounded conditional p-th moment, we have to resort to the Rosenthal inequality provided in Proposition 1.4.1. After applying the conditional Markov inequality which yields

$$\mathbb{P}\left(|\epsilon_{N_2}| \geq \frac{\delta}{2}\right) = \mathbb{E}\left[\mathbb{P}\left(|\epsilon_{N_2}| \geq \frac{\delta}{2}\Big| Z\right)\right] \leq \mathbb{E}\left[\mathbb{E}\left[|\epsilon_{N_2}|^p \mid Z\right] \cdot \left(\frac{2}{\delta}\right)^p\right]$$

we apply the conditional Rosenthal inequality (cf. Proposition 1.4.1) either for $1 < p \leq 2$ or for $p > 2$. In the first case, we get

$$\mathbb{E}\left[\mathbb{E}\left[|\epsilon_{N_2}|^p \mid Z\right] \cdot \left(\frac{2}{\delta}\right)^p\right] \leq C_p \cdot \mathbb{E}\left[\mathbb{E}[|W_j|^p \mid Z]\right] \cdot N_2^{-p+1} = C_p \cdot \mathbb{E}[|W_j|^p] \cdot N_2^{-p+1},$$

for $W_j = V_j - X$ and C_p depending on p. In the second case, we get (completely analogous to the proof of Proposition 3.2.6)

$$\mathbb{E}\left[|\epsilon_{N_2}|^p \mid Z\right] \cdot \left(\frac{2}{\delta}\right)^p \leq C_p \cdot N_2^{-(p/2)} \cdot \left(N_2^{-p/2+1} \cdot \mathbb{E}[|W_j|^p \mid Z] + \mathbb{E}[W_j^2 \mid Z]\right) \cdot \left(\frac{2}{\delta}\right)^p$$

and thus

$$\mathbb{E}\left[\mathbb{E}\left[|\epsilon_{N_2}|^p \mid Z\right] \cdot \left(\frac{2}{\delta}\right)^p\right] \le C_p \cdot N_2^{-(p/2)} \cdot \left(N_2^{-p/2+1} \cdot \mathbb{E}[|W_j|^p] + \mathbb{E}[W_j^2]\right) \cdot \left(\frac{2}{\delta}\right)^p$$

which proves the first and the second claim. $\qquad\square$

3.3 Inner and Outer Limit and a First Result on a.s. Convergence

In the subsequent paragraphs, we derive two different representations for the error term $\Delta\gamma_{N_1,N_2} = \bar{\gamma}_{N_1,N_2} - \gamma$. These different representations allow slightly different proof techniques in the following and also lead to a first result on a.s. convergence.

Inner and Outer Limit

We start with the *outer representation*

$$\Delta\gamma_{N_1,N_2} = \bar{\gamma}_{N_1,N_2} - \gamma = \bar{\gamma}_{N_1,N_2} - \mathbb{E}[G(X + \epsilon_{N_2})] + \mathbb{E}[G(X + \epsilon_{N_2})] - \mathbb{E}[G(X)]$$

$$= \left(\frac{1}{N_1}\sum_{i=1}^{N_1} G(X_i + \epsilon_{i,N_2}) - \mathbb{E}[G(X + \epsilon_{N_2})]\right) + \left(\mathbb{E}[G(X + \epsilon_{N_2})] - \mathbb{E}[G(X)]\right).$$

$$= \left(\frac{1}{N_1}\sum_{i=1}^{N_1} G(X_i + \epsilon_{i,N_2}) - \mathbb{E}[G(X + \epsilon_{N_2})]\right) + B_{N_2}, \tag{3.1}$$

where $B_{N_2} = \mathbb{E}[G(X + \epsilon_{N_2})] - \mathbb{E}[G(X)]$ represents the bias. For the *inner representation* we proceed as follows

$$\Delta\gamma_{N_1,N_2} = \frac{1}{N_1}\sum_{i=1}^{N_1}\left(G(X_i + \epsilon_{i,N_2}) - G(X_i) + G(X_i)\right) - \mathbb{E}[G(X)]$$

$$= \left(\frac{1}{N_1}\sum_{i=1}^{N_1}\left(G(X_i + \epsilon_{i,N_2}) - G(X_i)\right) + \frac{1}{N_1}\sum_{i=1}^{N_1} G(X_i) - \mathbb{E}[G(X)]\right)$$

$$= \left(\frac{1}{N_1}\sum_{i=1}^{N_1} G(X_i + \epsilon_{i,N_2}) - G(X_i) - B_{N_2}\right) + B_{N_2} + \left(\frac{1}{N_1}\sum_{i=1}^{N_1} G(X_i) - \mathbb{E}[G(X)]\right).$$

$$\tag{3.2}$$

Based on these representations, we consider the two 'special' cases, when either $N_1 = \infty$ or $N_2 = \infty$ and define the corresponding limiting quantities (outer and inner limit):

$$\Delta \gamma_{\infty, N_2} := \lim_{N_1 \to \infty} \Delta \gamma_{N_1, N_2}, \qquad \text{(outer limit)},$$

$$\Delta \gamma_{N_1, \infty} := \lim_{N_2 \to \infty} \Delta \gamma_{N_1, N_2}, \qquad \text{(inner limit)}.$$

The inner limit corresponds to the situation that we can observe X_i exactly, while we note that the outer limit collapses to a real number (and not a random variable anymore). The following Proposition 3.3.1 shows that both quantities are well defined under weak assumptions.

Proposition 3.3.1.
Suppose Assumptions (A) *and* (A0) *hold. Then both limits* $\Delta \gamma_{\infty, N_2}$ *(outer limit) and* $\Delta \gamma_{N_1, \infty}$ *(inner limit) are well-defined in an almost sure sense and it holds*

$$\Delta \gamma_{\infty, N_2} = \mathbb{E}[G(X + \epsilon_{N_2})] - \mathbb{E}[G(X)] = B_{N_2},$$

$$\Delta \gamma_{N_1, \infty} = \frac{1}{N_1} \sum_{i=1}^{N_1} G(X_i) - \mathbb{E}[G(X)].$$

The outer limit also exists in an L^1-*sense.*

Proof.
For fixed N_2, under Assumption (A), the SLLN holds for the iid sequence $\{G(X_i + \epsilon_{i, N_2})\}_{i \in \mathbb{N}}$, i.e.

$$\frac{1}{N_1} \sum_{i=1}^{N_1} G(X_i + \epsilon_{i, N_2}) \xrightarrow[N_1 \to \infty]{\text{a.s.}} \mathbb{E}[G(X + \epsilon_{N_2})]$$

and

$$\frac{1}{N_1} \sum_{i=1}^{N_1} G(X_i + \epsilon_{i, N_2}) \xrightarrow[N_2 \to \infty]{L^1} \mathbb{E}[G(X + \epsilon_{N_2})],$$

i.e. the limit exists both in an almost sure sense as well as in the L^1-sense, see e.g. Resnick (2005), p. 236, Exercise 8. Using representation (3.1), this proves the first equation.

Similarly, under Assumptions (A) and (A0), for fixed N_1, for each $i = 1, \ldots, N_1$, $\{\epsilon_{N_2,i}\}_{N_2 \in \mathbb{N}}$ converges to 0 almost surely by Proposition 1.4.1. Hence, we have for fixed N_1:

$$\frac{1}{N_1} \sum_{i=1}^{N_1} G(X_i + \epsilon_{i,N_2}) - G(X_i) \xrightarrow[N_2 \to \infty]{\text{a.s.}} 0.$$

Unfortunately, for convergence in L^1, additional assumptions on G like boundedness or Lipschitz continuity would be necessary. Representation (3.2), second line, yields the second statement.   □

From this result we can observe that the rate of convergence is closely linked to corresponding SLLN rates of convergence and to the order of the bias B_{N_2}.

A First Result on a.s. Convergence

From Proposition 3.3.1, we can observe that under Assumptions (A) and (A0) we have almost surely:

$$\lim_{N_1 \to \infty} \lim_{N_2 \to \infty} \Delta \gamma_{N_1, N_2} = \lim_{N_1 \to \infty} \Delta \gamma_{N_1, \infty} = 0,$$

since the law of large numbers holds for $\{G(X_i)\}_{i \in \mathbb{N}}$. Further assuming (VB) holds, we obtain almost surely:

$$\lim_{N_1 \to \infty} \lim_{N_2 \to \infty} \gamma_{N_1, N_2} = \gamma = \lim_{N_2 \to \infty} \lim_{N_1 \to \infty} \gamma_{N_1, N_2}.$$

Moreover, we can derive the subsequent Corollary 3.3.2 which significantly extends the results in Rainforth et al. (2018), Theorem 2, while making similar assumptions.

Corollary 3.3.2.
Suppose Assumptions (A), (A0) and (VB) hold. Then, there exists a function $n_1 : \mathbb{N} \to \mathbb{N}$ such that $\Delta \gamma_{n_1(N), N}$ converges to 0 almost surely. Further, for each $\delta > 0$ there exists $\Omega_\delta \in \mathcal{F}$ with $\mathbb{P}(\Omega_\delta) \geq 1 - \delta$ and a function $n_2 : \mathbb{N} \to \mathbb{N}$ (depending on δ) such that $\Delta \gamma_{N, n_2(N)}$ converges to 0 almost surely on Ω_δ.

Proof.

"$\Delta\gamma_{n_1(N),N}$": Let $N_2 = N$ be fix. Due to Assumption (A), the SLLN applies to the iid sequence $\{G(X_i + \epsilon_{i,N_2})\}_{i\in\mathbb{N}}$ and we especially have, see Resnick (2005), p. 236, Exercise 8,

$$\left\|\frac{1}{N_1}\sum_{i=1}^{N_1} G(X_i + \epsilon_{i,N_2}) - \mathbb{E}[G(X + \epsilon_{N_2})]\right\|_{L^1} \to 0 \quad \text{for } N_1 \to \infty.$$

Thus, for $N_2 = N$, choose $n_1(N) = N_1$ such that this norm is smaller than $1/N^2$. By this choice we have for fixed $\varepsilon > 0$

$$\sum_{N\in\mathbb{N}} \mathbb{P}\left(\left|\frac{1}{n_1(N)}\sum_{i=1}^{n_1(N)} G(X_i + \epsilon_{N,i}) - \mathbb{E}[G(X + \epsilon_N)]\right| > \varepsilon\right) \le \sum_{N\in\mathbb{N}} \frac{1}{\varepsilon N^2} < \infty$$

using Markov's inequality. Since this yields complete convergence, this implies almost sure convergence and the statement is proved by using representation (3.1).

"$\Delta\gamma_{N,n_2(N)}$": Due to Proposition 3.3.1 the inner limit exists a.s., hence we can apply the same ideas (i.e. Egorov's theorem) as in the proof of Rainforth et al. (2018), Theorem 2. □

The results in Corollary 3.3.2 extend Rainforth et al. (2018), Theorem 2, in two ways: First, we provide a similar result for $\Delta\gamma_{N,n_2(N)}$ as for $\Delta\gamma_{n_1(N),N}$ in Rainforth et al. (2018), Theorem 2, and second, we get rid of the constraining set A_δ for $\Delta\gamma_{n_1(N),N}$ from Rainforth et al. (2018), Theorem 2. The latter result is due to the fact that the SLLN not only provides convergence in the almost sure sense, but also convergence in mean—a fact which seems to be overlooked by previous work on nested simulations.

Like other authors before, we note that the results and the proof of Corollary 3.3.2 indicate that if N_1 goes to ∞ too slowly (compared to N_2), almost sure convergence might not hold.

Finally, in our opinion, the results of Corollary 3.3.2 are still not satisfying: unlike existing results on L^2-convergence, no *explicit* budget sequence is derived which guarantees almost sure convergence. In the next section, we will therefore consider stronger results, however, at the expense of slightly stronger assumptions.

3.4   Main Results on a.s. Convergence

We are now ready to formulate our main results concerning almost sure convergence and corresponding convergence rates. For these stronger results, we need to make a few additional weak assumptions concerning higher moments.

Almost Sure Convergence without Convergence Rates

Theorem 3.4.1.
Let $p \geq 2$ and let Assumption (A1,p) hold. Further, let $\{(N_1(N), N_2(N))\}$ be an arbitrary budget sequence and let $m_p(N) := \mathbb{E}[|G(X + \epsilon_{N_2(N)})|^p]$. Then, for each $t > 0$ there exists an $\bar{N} \in \mathbb{N}$ such that

$$\forall N \geq \bar{N}: \quad \mathbb{P}\left(\left|\Delta\gamma_{N_1(N),N_2(N)}\right| > t\right) \leq 12^{p+1} \cdot t^{-p} \cdot m_p(N) \cdot N_1(N)^{-p+1},$$

and thus the sequence $\left\{\bar{\gamma}_{N_1(N),N_2(N)}\right\}$ converges in probability to γ. If the budget sequence satisfies $\sum_{N=1}^{\infty} N_1(N)^{-p+1} < \infty$, then

$$\Delta\gamma_{N_1(N),N_2(N)} \xrightarrow[N \to \infty]{a.s.} 0.$$

Proof.
The first statement follows directly from Lemma 3.4.3, given below, and the fact that $\{(N_1(N), N_2(N))\}$ is a budget sequence. Since $m_p(N)$ is bounded according to Assumption (A1,p), and since $N_1(N)^{-p+1}$ goes to 0 for each budget sequence (as $p \geq 2$), we immediately obtain convergence in probability. Almost sure convergence then follows directly from complete convergence which holds if the sufficient condition $\sum_{N=1}^{\infty} N_1(N)^{-p+1} < \infty$ is satisfied. $\square$

We especially note that the very weak (uniform) integrability assumption on $G(X + \epsilon_{N_2})$ is already sufficient for the stated convergence in probability for *arbitrary* budget sequences; there is no need for further assumptions concerning the bias or the structure of G.

Remark 3.4.2.
While for $p = 2$ the sufficient condition $\sum_{N=1}^{\infty} N_1(N)^{-p+1} < \infty$ in Theorem 3.4.1 is not satisfied for any budget sequence, it can be easily satisfied for $p > 2$: A comparison to the convergent series $\sum_{N=1}^{\infty} N^{\lambda}$ for $\lambda < -1$ shows that it is sufficient

to require $N_1(N) \geq N^{1-\alpha}$ with $0 < \alpha < \frac{p-2}{p-1}$. We thus observe again that almost sure convergence seems to require more outer than inner scenarios, at least for small values of p. The more moments exist, the more freedom consists in the choice of the budget sequence.

Lemma 3.4.3.
Let $p \geq 2$, let Assumption (A1,p) hold, and let $m_p(N_2) := \mathbb{E}[|G(X + \epsilon_{N_2})|^p]$. Then, for each $t > 0$ there exist $\bar{N}_1, \bar{N}_2 \in \mathbb{N}$ such that

$$\forall N_1 \geq \bar{N}_1, \forall N_2 \geq \bar{N}_2 : \quad \mathbb{P}\left(\left|\Delta\gamma_{N_1,N_2}\right| > t\right) \leq 6 \cdot (\eta/6)^{-p} \cdot t^{-p} \cdot m_p(N_2) \cdot N_1^{-p+1},$$

where $\eta = p/(p+2)$.

Proof.
Let $\tilde{t} = \frac{t}{3}$. By Assumption (A1,p), it follows from Proposition 3.1.1(4) that there exists $\bar{N}_2 \in \mathbb{N}$ such that for all $N_2 \geq \bar{N}_2$ we have $\left|\mathbb{E}[G(X + \epsilon_{N_2})] - \mathbb{E}[G(X)]\right| \leq \tilde{t}$. For $N_2 \geq \bar{N}_2$, we thus obtain for $\mathbb{P}\left(\left|\Delta\gamma_{N_1,N_2}\right| > 2\tilde{t}\right)$:

$$\mathbb{P}\left(\left|\frac{1}{N_1}\sum_{i=1}^{N_1} G(X_i + \epsilon_{i,N_2}) - \mathbb{E}[G(X)]\right| > 2\tilde{t}\right) \leq \mathbb{P}\left(\left|\frac{1}{N_1}\sum_{i=1}^{N_1} G(X_i + \epsilon_{i,N_2}) - \mathbb{E}[G(X + \epsilon_{N_2})]\right| > \tilde{t}\right).$$

Therefore, it remains to show

$$\mathbb{P}\left(\left|\frac{1}{N_1}\sum_{i=1}^{N_1} G(X_i + \epsilon_{i,N_2}) - \mathbb{E}[G(X + \epsilon_{N_2})]\right| > \tilde{t}\right) \leq 6 \cdot (\eta/2)^{-p} \cdot \tilde{t}^{-p} \cdot m_p(N_2) \cdot N_1^{-p+1},$$

for sufficiently large N_1. For this purpose, we note that $\{G(X_i + \epsilon_{i,N_2})\}_{i \in \mathbb{N}}$ are iid and have finite p-th moment according to Assumption (A1,p) (if necessary, we have to increase $\bar{N}_2$ appropriately). Hence, we can apply Corollary 2.2.2 with $Y_i = G(X_i + \epsilon_{i,N_2}) - \mathbb{E}[G(X + \epsilon_{N_2})]$ and obtain for fixed N_2 the existence of some $\tilde{N}_1(N_2, \tilde{t})$ such that for all $N_1 \geq \tilde{N}_1(N_2, \tilde{t})$

$$\mathbb{P}\left(\left|\frac{1}{N_1}\sum_{i=1}^{N_1} G(X_i + \epsilon_{i,N_2}) - \mathbb{E}[G(X + \epsilon_{N_2})]\right| > \tilde{t}\right) \leq 6 \cdot (\eta/2)^{-p} \cdot \tilde{t}^{-p} \cdot m_p(N_2) \cdot N_1^{-p+1}$$

holds. Note that we have used that $\mathbb{E}[|Y_i|^p] \leq 2^p \|G(X + \epsilon_{N_2})\|_{L^p}^p$. Since all $m_p(N_2)$ are bounded above by some constant according to Assumption (A1,p),

Corollary 2.2.2 actually yields that $\bar{N}_1(N_2, \tilde{t})$ can be chosen independently of N_2, which proves the claim. $\qquad\square$

Almost Sure Convergence with Convergence Rates

For the consideration of convergence rates, it is necessary to assume that the bias vanishes with a given order. More details on the asymptotic behaviour of the bias for a variety of function classes is given in Section 3.2. Based on this vanishing bias assumption and a moment condition on $G(X + \epsilon_{N_2})$, cf. (A1,p), we obtain now optimal rates of convergence in probability and in an almost sure sense, i.e.:

Theorem 3.4.4.
Let $p \geq 2$, $0 < \beta_1 < 1/2$, $0 < \beta_2$ and suppose Assumptions (A1,p) and (VB,β) hold for $\beta = \beta_2$. Further, let $0 < r < 1$ and consider the budget sequence $N_1(N) = \lfloor N^{1-r} \rfloor$, $N_2(N) = \lfloor N^r \rfloor$ and let $m_p(N) := \mathbb{E}[|G(X + \epsilon_{N_2(N)})|^p]$. Then it holds:

(1.) If $0 < r \leq \beta_1/(\beta_1 + \beta_2)$, then the sequence $N_2(N)^{\beta_2} \cdot |\Delta \gamma_{N_1(N),N_2(N)}|$ converges to 0 in probability; more exactly, there exists an $\bar{N} \in \mathbb{N}$ such that:

$$\forall N \geq \bar{N}: \quad \mathbb{P}\left(N_2(N)^{\beta_2} \cdot |\Delta \gamma_{N_1(N),N_2(N)}| > t\right) \leq 12^{p+1} \cdot t^{-p} \cdot m_p(N) \cdot N_1(N)^{-p(1-\beta_1)+1}.$$

(2.) If $0 < r \leq \beta_1/(\beta_1 + \beta_2)$, and if further $\beta_1 < (p-2)/p$ and $r < \frac{p(1-\beta_1)-2}{p(1-\beta_1)-1}$, then the sequence $N_2(N)^{\beta_2} \cdot |\Delta \gamma_{N_1(N),N_2(N)}|$ converges to 0 almost surely.

Proof.
The first statement follows from Lemma 3.4.6, given below, and the fact that the budget sequence $\{(N_1(N), N_2(N))\}$ satisfies $N_1(N)^{\beta_1} \geq N_2(N)^{\beta_2}$ for all N by construction. The right-hand side in (1.) goes to 0 as we consider a budget sequence and as $-p(1-\beta_1) + 1 < 0$ holds for all $p \geq 2$ and $0 < \beta_1 < 1/2$.

For the second statement, we first note that the denominator of $\frac{p(1-\beta_1)-2}{p(1-\beta_1)-1}$ is positive for all $p \geq 2$ and $0 < \beta_1 < 1/2$ and that the nominator of $\frac{p(1-\beta_1)-2}{p(1-\beta_1)-1}$ is positive if and only if $\beta_1 < (p-2)/p$. Now, for each $0 < r < \frac{p(1-\beta_1)-2}{p(1-\beta_1)-1}$ it holds that $(1-r)(-p(1-\beta_1)+1) < -1$, which – using the first statement – yields complete convergence by the same line of arguments as in the proof of Theorem 3.4.1 and Remark 3.4.2. $\qquad\square$

Remark 3.4.5.
Let us elaborate a bit more on the results given in Theorem 3.4.4:

- *In Theorem 3.4.4(1), for given β_1 and β_2, we obtain the best asymptotic convergence rate $N_2(N)^{\beta_2} = N^{\beta_1\beta_2/(\beta_1+\beta_2)}$ for the largest possible r, i.e. by the choice $r = \beta_1/(\beta_1 + \beta_2)$. We note that this choice and the best rate is independent of the number of existing moments.*
- *Similarly, in Theorem 3.4.4(2), for given p, β_1 and β_2, the best possible rate is also obtained if r is chosen as large as possible (since we want to maximize $r \cdot \beta_2$).*
- *In essence, for both types of convergence, it holds that the more budget can be shifted to the inner scenarios, the better the rate which can be achieved. This is mainly because the bias usually has a better rate than the law of large numbers in a pure Monte Carlo setting.*
- *We note that in both cases (for given p which is determined by the existing moments and for given β_2 which is determined by the bias) we have a degree of freedom in the choice of β_1. While for convergence in probability, it is easily seen that β_1 should always be chosen as large as possible (i.e. almost $1/2$, to allow for large α), the situation is more subtle for almost sure convergence, where an (easy one-dimensional) optimization over the free parameter β_1 has to be carried out*
- *For convergence in probability, the best rate (which cannot be achieved, but one can get arbitrarily close) is hence given by $N^{\beta_2/(1+2\beta_2)}$ (for β_1 almost $1/2$). For the common case where (almost) $\beta_2 = 1/2$, this yields a convergence rate of (almost) $N^{1/4}$, while under strong assumptions as e.g. made in Hong and Juneja (2009), Lee (1998), Liu et al. (2022) or Section 3.2 one (almost) has $\beta_2 = 1$ and thus, an asymptotic rate of (almost) $N^{1/3}$. In this way, we supplement the rates originally obtained in Hong and Juneja (2009) for L^2-convergence (and thus in probability) which are obtained under strong assumptions by (slightly weaker) rates under much less restrictive assumptions. Since the best rate is given by (almost) $N^{\beta_2/(1+2\beta_2)}$ (for β_1 almost $1/2$), it becomes obvious that even if the bias vanishes arbitrarily fast, no order better than $-1/2$ can be reached by the uniform estimator for convergence in probability. Taking into account Proposition 3.2.8, we see that this rate is generically achieved for the large loss probability if Z takes only finitely many values.*
- *Concerning almost sure convergence, as noted in Remark 3.4.2, we also note here that for $p = 2$ the condition on β_1 in Theorem 3.4.4(2) cannot be satisfied. We further note that it becomes redundant for $p \geq 4$. From Theorem 3.4.4(2) it now becomes clear that as soon as $p > 2$ holds, there exist budget sequences which yield a.s. convergence; albeit with potentially low rates. For specific examples,*

let us refer to Table 3.1 which shows the optimal choice of β_1 and α for given p and β_2. We can observe that the optimal rate $N^{1/3}$ is already obtained by the choice β_1 equals (almost) $1/2$ for $p \geq 5$ and $\beta_2 = 1$. In contrast, if $\beta_2 = 1/2$, the optimal rate of $N^{1/4}$ is only obtained for $p \geq 6$.

Based on a variation of the Fuk-Nagaev inequality, cf. Corollary 2.2.3, Lemma 3.4.6 provides the needed probability inequality for the approximation error $\Delta\gamma_{N_1,N_2}$ to obtain in the above Theorem 3.4.4 the corresponding rate of convergence in probability and to get according to the complete convergence also the desired almost sure convergence. Again, the speed of convergence depends crucially on the underlying order of the bias, i.e. (VB,β).

Table 3.1 Optimal rates for almost sure convergence for given p and β_2

(p, β_2)	β_1	r	$N^{r \cdot \beta_2}$
$(3, \frac{1}{2})$	$\frac{1}{5}$	$\frac{2}{7}$	$\frac{1}{7}$
$(4, \frac{1}{2})$	$\frac{1}{3}$	$\frac{2}{5}$	$\frac{1}{5}$
$(5, \frac{1}{2})$	$\frac{3}{7}$	$\frac{6}{13}$	$\frac{3}{13}$
$(6, \frac{1}{2})$	$\frac{1}{2}$	$\frac{1}{2}$	$\frac{1}{4}$
$(3, 1)$	$\frac{1}{4}$	$\frac{1}{3}$	$\frac{1}{3}$
$(4, 1)$	$\frac{2}{5}$	$\frac{2}{7}$	$\frac{2}{7}$
$(5, 1)$	$\frac{1}{2}$	$\frac{1}{3}$	$\frac{1}{3}$

Lemma 3.4.6.

Let $p \geq 2$, $0 < \beta_1 < 1/2$, $0 < \beta_2$, suppose Assumptions (A1,p) and (VB,β) hold for $\beta = \beta_2$ and let $m_p(N_2) := \mathbb{E}[|G(X + \epsilon_{N_2})|^p]$. Then, for each $t > 0$ there exists $\bar{N}_1$, $\bar{N}_2 \in \mathbb{N}$ such that for all $N_1 \geq \bar{N}_1$ and for all $N_2 \geq \bar{N}_2$ with $N_1^{\beta_1} \geq N_2^{\beta_2}$

$$\mathbb{P}\left(N_2^{\beta_2} \cdot \left|\Delta\gamma_{N_1,N_2}\right| > t\right) \leq 6 \cdot (\eta/6)^{-p} \cdot t^{-p} \cdot m_p(N_2) \cdot N_1^{-p(1-\beta_1)+1},$$

applies, with $\eta = p/(p+2)$.

Proof.
Let $\tilde{t} = \frac{t}{3}$. By Assumption (VB,β), it follows that there exists $\bar{N}_2 \in \mathbb{N}$ such that for all $N_2 \geq \bar{N}_2$ we have $\left|\mathbb{E}[G(X + \epsilon_{N_2})] - \mathbb{E}[G(X)]\right| \leq \tilde{t}/N_2^{\beta_2}$. Let $\eta_{i,N_2} := G(X_i + \epsilon_{i,N_2}) - \mathbb{E}[G(X)]$ and $\bar{\eta}_{i,N_2} := G(X_i + \epsilon_{i,N_2}) - \mathbb{E}[G(X + \epsilon_{N_2})]$, then for $N_2 \geq \bar{N}_2$, we thus obtain for $\mathbb{P}\left(N_2^{\beta_2}\left|\Delta\gamma_{N_1,N_2}\right| > 2\tilde{t}\right)$:

$$\mathbb{P}\left(N_2^{\beta_2}\left|\frac{1}{N_1}\sum_{i=1}^{N_1}\eta_{i,N_2}\right| > 2\tilde{t}\right) \leq \mathbb{P}\left(N_2^{\beta_2}\left|\frac{1}{N_1}\sum_{i=1}^{N_1}\bar{\eta}_{i,N_2}\right| > \tilde{t}\right).$$

If we assume $N_1^{\beta_1} \geq N_2^{\beta_2}$, it remains to show

$$\mathbb{P}\left(N_1^{\beta_1}\left|\frac{1}{N_1}\sum_{i=1}^{N_1}\bar{\eta}_{i,N_2}\right| > \tilde{t}\right) \leq 6 \cdot (\eta/2)^{-p} \cdot \tilde{t}^{-p} \cdot m_p(N_2)N_1^{-p(1-\beta_1)+1},$$

for sufficiently large N_1. For this purpose, we note that $\{G(X_i + \epsilon_{i,N_2})\}_{i\in\mathbb{N}}$ are iid and have finite p-th moment according to Assumption (A1,p) (if necessary, we have to increase $\bar{N}_2$ appropriately). Hence, we can apply Corollary 2.2.3 with $Y_i = G(X_i + \epsilon_{i,N_2}) - \mathbb{E}[G(X + \epsilon_{N_2})]$ and obtain for fixed N_2 the existence of some $\bar{N}_1(N_2, \tilde{t})$ such that

$$\forall N_1 \geq \bar{N}_1(N_2, \tilde{t}): \quad \mathbb{P}\left(N_1^{\beta_1}\left|\frac{1}{N_1}\sum_{i=1}^{N_1}\bar{\eta}_{i,N_2}\right| > \tilde{t}\right) \leq 6\cdot(\eta/2)^{-p}\cdot\tilde{t}^{-p}\cdot m_p(N_2)\cdot N_1^{-p(1-\beta_1)+1}.$$

Note that we have used $\mathbb{E}[|Y_i|^p] \leq 2^p\|G(X + \epsilon_{N_2})\|_{L^p}^p$. Since all $m_p(N_2)$ are bounded above by some constant according to Assumption (A1,p), Corollary 2.2.3 actually yields that $\bar{N}_1(N_2, \tilde{t})$ can be chosen independently of N_2 which proves the claim. $\square$

Almost Sure Convergence of Quantile-Based Estimators

4

Here, we investigate in more detail the quantile-based problem, cf. (1.11), especially the so far unexamined almost sure convergence, cf. Table 1.2. Note, that the whole chapter grounds on the results of Klein and Werner (2023c). The research question at hand is similar to the previous one but now applied to the empirical quantile estimator $\widehat{q}_{\alpha,N_1}^{X+\epsilon_{N_2}}$. Hence, we are interested in the growth conditions on N_1 and N_2 which enable an uniform almost sure convergence of $\widehat{q}_{\alpha,N_1}^{X+\epsilon_{N_2}}$ towards the true quantile q_α^X. On top of this, we also address a so far unanswered question and derive promising rates of convergence in an almost sure framework.

The Chapter is divided into the following sections: In Section 4.1 we discuss the necessary assumptions which are needed for our main results. Here, we focus on the minimum requirements, especially on the tail behaviour of ϵ_{N_2}. Furthermore, we introduce in accordance to the previous vanishing bias discussion the so-called vanishing quantile. In Section 4.2 we derive explicit vanishing quantile rates for our relatively weak assumptions (cf. Section 4.1) and provide a remarkable rate of $-1/2$. Similar to the prove of Gordy and Juneja (2010) (cf. Proof of Equation (27)) and based on their assumptions (cf. Assumption 1) we are able to carry their large loss rate over to the problem at hand which yields an optimal rate of -1. In Section 4.3 we present, first, a representation of the underlying approximation error, i.e. $\left|\widehat{q}_{\alpha,N_1}^{X+\epsilon_{N_2}} - q_\alpha^X\right|$ and second, we provide our novel results, Theorem 4.3.1 and Theorem 4.3.5. Both state the convergence in probability as well as the almost sure convergence of $\widehat{q}_{\alpha,N_1}^{X+\epsilon_{N_2}}$. Thereby, Theorem 4.3.1 states the strong consistency of the estimator, whereby Theorem 4.3.5 additionally considers a rate factor N_2^β and thus derives promising rates for the almost sure convergence. It should be noted that our numerical investigations on these results are summarized in Chapter 6.

© The Author(s), under exclusive license to Springer Fachmedien Wiesbaden GmbH, part of Springer Nature 2024
M. Klein, *Nested Simulations: Theory and Application*, Mathematische Optimierung und Wirtschaftsmathematik | Mathematical Optimization and Economathematics, https://doi.org/10.1007/978-3-658-43853-1_4

4.1   Assumptions

Our main quantities in this setup are the random variables X, ϵ_{N_2}, q_α^X and $q_\alpha^{X+\epsilon_{N_2}}$. Hence, instead of considering expectations, we are now interested in quantiles and therefore the SCR relevant case. Again, as in the previous section we make a very weak basic assumption on V, i.e.

$$V \in L^1(\Omega, \mathcal{F}, \mathbb{P}). \tag{B}$$

This implies $X \in L^1(\Omega, \mathcal{F}, \mathbb{F})$ and especially $\epsilon_{N_2} \in L^1(\Omega, \mathcal{F}, \mathbb{P})$ for each N_2, cf. Proposition 1.4.1.

Non-flatness and Integrability Assumptions

We cannot expect any convergence result, if the underlying distribution function F_X is locally flat at the respective quantile q_α^X since the naive estimation of the Value-at-Risk of a nested simulation is based on the estimation of the corresponding quantile $q_\alpha^{X+\epsilon_{N_2}}$ of a distorted distribution $F_{X+\epsilon_{N_2}}$. Hence, an obvious minimum requirement is that the sequence of standard estimators $\{\widehat{q}_{\alpha,N_1}^X\}$ will at least converge almost surely to q_α^X. Therefore, we make the following weak assumption:

The distribution function F_X is differentiable at q_α^X and $F'_X(q_\alpha^X) > 0$ holds. (B0)

As a consequence of a simple Taylor expansion of order one this assumption implies directly the so-called 'non-flatness' condition and thus that (the slope)

$$\delta(\epsilon, X) := \min \left\{ \alpha - F_X(q_\alpha^X - \epsilon), F_X(q_\alpha^X + \epsilon) - \alpha \right\} \tag{NF}$$

is strictly positive for all $\epsilon > 0$. As already mentioned in the introduction, with (NF) we immediately obtain that the sequence of standard estimators $\left\{ \widehat{q}_{\alpha,N_1}^X \right\}$ converges almost surely to q_α^X (see for instance Theorem on page 75 in Serfling (1980)). The obvious question is now, which additional assumptions are needed to ensure an almost sure convergence of $\left\{ \widehat{q}_{\alpha,N_1}^{X+\epsilon_{N_2}} \right\}$, based on a nested Monte Carlo simulation towards q_α^X. To this end more knowledge about the tail behaviour of ϵ_{N_2} is essential.

Hence, we make the following assumption on the moments of ϵ_{N_2} (more precisely on V), i.e.

$$\|V\|_{L^p} < \infty, \tag{B1,p}$$

respectively $V \in L^p(\Omega, \mathcal{F}, \mathbb{P})$. These are the core assumptions to obtain in the following almost sure convergence results and their corresponding order.

Assumptions on the Quantile Difference

Crucial in our considerations will be the convergence behaviour of the *quantile difference* denoted by $Q_{N_2} := q_\alpha^{X+\epsilon_{N_2}} - q_\alpha^X$, i.e. the difference between the distorted and the true quantile. Thus, a natural assumption would be that the quantile difference vanishes asymptotically, i.e.:

$$q_\alpha^{X+\epsilon_{N_2}} \xrightarrow[N_2 \to \infty]{} q_\alpha^X. \tag{VQ}$$

We summarize this characteristic under the heading *vanishing quantile*. But, in order to obtain later on also asymptotic convergence rates, we must also consider a rate factor and thus

$$N_2^\beta \cdot \left| q_\alpha^{X+\epsilon_{N_2}} - q_\alpha^X \right| \xrightarrow[N_2 \to \infty]{} 0, \tag{VQ,β}$$

for some $\beta \geq 0$. As for the moment-based case we note that here (VQ) coincides also with (VQ,β) for $\beta = 0$. In the following, we investigate two different assumption sets and prove the corresponding order in (VQ,β).

4.2   Quantile Difference Considerations

In the upcoming paragraphs, we consider the vanishing quantile problem and derive two different rates β. The first is attainable under the already introduced weak assumptions for e.g. non differentiable functions, whereas the second relies on the results of Gordy and Juneja (2010) under much stricter assumptions. These will be addressed and discussed in the respective paragraph. Note, that this is a rather interesting question since it describes a crucial point in further analyses. Overall, we prove that (VQ,β) holds for $0 \leq \beta < 1/2$ under weak assumptions, furthermore we show for $0 \leq \beta < 1$ that the large loss bias rate (VB,β) (cf. Gordy and Juneja (2010)) can be used to obtain (VQ,β).

Function Class Satisfying (VQ,β) for $0 \leq \beta < 1/2$

Proposition 4.2.1.
Let Assumptions (B0) *and* (B1,p) *hold with* $p \geq 2$. *Then, for* $0 \leq \beta <$ min $\left\{ \frac{1}{2}, \frac{p-1}{p+1} \right\}$ *and* $\epsilon > 0$, *there exists an* $\bar{N}_2 \in \mathbb{N}$ *such that*

$$\forall N_2 \geq \bar{N}_2 : \quad \left| q_\alpha^{X+\epsilon_{N_2}} - q_\alpha^X \right| \leq \frac{\epsilon}{N_2^\beta}.$$

Proof.
We first proof the statement for $\epsilon = 1$. It is sufficient to show

$$F_{X+\epsilon_{N_2}} \left(q_\alpha^X + \frac{1}{N_2^\beta} \right) > \alpha. \quad \text{and} \quad F_{X+\epsilon_{N_2}} \left(q_\alpha^X - \frac{1}{N_2^\beta} \right) < \alpha.$$

Using Petrov (1975), Lemma 3, yields for all N_2 and $t > 0$

$$F_{X+\epsilon_{N_2}} \left(q_\alpha^X + \frac{1}{N_2} \right) \geq F_X \left(q_\alpha^X + \frac{1}{N_2^\beta} - t \right) - \mathbb{P} \left(|\epsilon_{N_2}| \geq t \right).$$

We set $t = 1/(2N_2^\beta)$ and obtain for all N_2:

$$F_{X+\epsilon_{N_2}} \left(q_\alpha^X + \frac{1}{N_2} \right) \geq F_X \left(q_\alpha^X + \frac{1}{N_2^\beta} - \frac{1}{2N_2^\beta} \right) - \mathbb{P} \left(|\epsilon_{N_2}| \geq \frac{1}{2N_2^\beta} \right).$$

According to (B1,p), we obtain with Proposition 1.4.1 that $||\epsilon_{N_2}||_{L^p} < \infty$ holds. Hence, Corollary 2.2.3 with $t = 1/2$ yields for $N_2 \geq \bar{N}_2$

$$F_{X+\epsilon_{N_2}} \left(q_\alpha^X + \frac{1}{N_2} \right) \geq F_X \left(q_\alpha^X + \frac{1}{2N_2^\beta} \right) - d_p \cdot N_2^{-p(1-\beta)+1},$$

whereby $d_p := 3 \cdot C_p \cdot 2^p$. Applying the definition of the derivative of a real function we get for all N_2:

$$F_X\left(q_\alpha^X + \frac{1}{2N_2^\beta}\right) = F_X(q_\alpha^X) + F_X'(q_\alpha^X) \cdot \frac{1}{2N_2^\beta} + R(N_2) \cdot \frac{1}{2N_2^\beta}$$

with residual term $R(N_2) \to 0$ for $N_2 \to \infty$. Hence, if N_2 is large enough, we have $|R(N_2)| < (1/2)F_X'(q_\alpha^X)$. Thus, for $N_2 \geq \widehat{N_2}$ with sufficiently large $\widehat{N_2}$ (and $\widehat{N_2} \geq \bar{N_2}$) we obtain

$$F_{X+\epsilon_{N_2}}\left(q_\alpha^X + \frac{1}{N_2}\right) \geq F_X(q_\alpha^X) + F_X'(q_\alpha^X) \cdot \frac{1}{4N_2^\beta} - d_p \cdot N_2^{-p(1-\beta)+1}.$$

Since $F_X(q_\alpha^X) = \alpha$ holds, the claim is proved, if $F_X'(q_\alpha^X) \cdot \frac{1}{4N_2^\beta} - d_p \cdot N_2^{-p(1-\beta)+1} > 0$ holds for sufficiently large N_2. This is equivalent to $F_X'(q_\alpha^X) > 4d_p N_2^{\beta(p+1)-(p-1)}$ which can be easily satisfied for large enough N_2 if $\beta < \frac{p-1}{p+1}$. The remaining claim follows along similar lines of arguments. Note, that the statement for general $\epsilon > 0$ now follows straightforwardly since the power β can be enlarged to

$$\beta' := \frac{1}{2}\left(\beta + \min\left\{\frac{1}{2}, \frac{p-1}{p+1}\right\}\right).$$

$\square$

Remark 4.2.2.
We note that only for $p \geq 3$ the better convergence order of (almost) $-1/2$ in N_2 will be reached and that for $p = 2$ the convergence order is only $-1/3$. Hence, the number of existing moments in (B1,p) plays a crucial role. Let us point out that the order of (almost) $-1/2$ can only be obtained due to the fact that ϵ_{N_2} is an average of independent variables which has an immediate impact on the tail behaviour of ϵ_{N_2}. If this is neglected and only the existence of moments is taken into account, Example 4.2.3, part (i) shows that no order better than $-1/3$ can be expected.

Since the Fuk-Nagaev inequality describes the best obtainable rate for the law of large numbers, we believe that the rate in Proposition 4.2.1 might only be improved if (together with the tail behaviour of ϵ_{N_2}) the condition $\mathbb{E}\left[\epsilon_{N_2}|X\right] = 0$ or further smoothness assumptions on F_X are taken into account. However, we were not able to improve the rate in this direction any further based on the so far made weak assumptions. But, note that the best rate one can hope for is -1, as part (ii) of Example 4.2.3 clearly indicates. This rate, obtained by Gordy and Juneja (2010), and its corresponding strict smoothness assumptions will be discussed in more detail in the following paragraph.

Example 4.2.3.

(i) Let $X \sim \mathcal{U}_{[0,1]}$ be uniform distributed on $[0, 1]$. Then, it holds $q_\alpha^X = \alpha$. Let further

$$\epsilon_{N_2} = \frac{1}{N_2^{1/3}} \left(\mathbb{1}_{X \in \left[1-\alpha-1/\left(2N_2^{1/3}\right)\right], 1-\alpha+1/\left(2N_2^{1/3}\right)} - \mathbb{1}_{X \in \left[\alpha-1/\left(2N_2^{1/3}\right), \alpha+1/\left(2N_2^{1/3}\right)\right]} \right)$$

for N_2 large enough. By construction we get

$$\left| q_\alpha^{X+\epsilon_{N_2}} - q_\alpha^X \right| \geq \frac{1}{2 \cdot N_2^{1/3}}.$$

(ii) Let $X \sim \mathcal{N}(0, 1)$ and $\epsilon_{N_2} \sim \mathcal{N}(0, 1/N_2)$ be independently normal distributed. Then, $q_\alpha^X = \Phi^{-1}(\alpha)$ and $q_\alpha^{X+\epsilon_{N_2}} = \Phi^{-1}(\alpha)\sqrt{1 + 1/N_2}$ applies. Here, the difference of these quantiles is clearly of order -1 in N_2.

Function Class Satisfying (VQ,β) for $0 \leq \beta < 1$

Example 4.2.3 part (ii) demonstrates that an order of -1 is possible under strong smoothness conditions. Hence, in the following we repeat – once again (cf. Chapter 1) – these strong assumptions of Gordy and Juneja (2010) but now in more detail, bring them in our context and summarize the (VQ,β) result for $0 \leq \beta < 1$ in Proposition 4.2.6. Note, that Liu et al. (2022) pick up this result and the corresponding assumption in their recent publication. However, they were not able to weaken the assumptions any further.

In general, Gordy and Juneja (2010) make assumptions on the behaviour of the joint density function $f_{(Z,V)}$ of (Z, V), like differentiability (twice differentiable) and that they are bounded. More technically, the behaviour of X and $\tilde{\epsilon}_{N_2} := \sqrt{N_2} \cdot \epsilon_{N_2}$ will be restricted.

Assumption 1. *(Gordy and Juneja (2010), Assumption 1)*
Let $f_{N_2}(x, y)$ denote the joint density of $(X, \tilde{\epsilon}_{N_2})$. We now suppose the following:

(i) The joint density $f_{N_2}(x, y)$ and its corresponding partial derivatives

$$\frac{\partial}{\partial x} f_{N_2}(x, y) \quad and \quad \frac{\partial^2}{\partial x^2} f_{N_2}(x, y)$$

exist for each N_2 and for all (x, y).

(ii) For $N_2 \geq 1$, there exist non-negative functions p_{0,N_2}, p_{1,N_2} and p_{2,N_2} such that

$$f_{N_2}(x, y) \leq p_{0,N_2}(y),$$

$$\left| \frac{\partial}{\partial x} f_{N_2}(x, y) \right| \leq p_{1,N_2}(y),$$

$$\left| \frac{\partial^2}{\partial x^2} f_{N_2}(x, y) \right| \leq p_{2,N_2}(y),$$

hold for all x, y, i.e. the joint density and its corresponding partial derivatives are bounded. Additionally,

$$\sup_{N_2} \int_{-\infty}^{\infty} |y|^r \, p_{i,N_2}(y) \, dy < \infty$$

is assumed for $i = 0, 1, 2$ and $0 \leq r \leq 4$.

In order to obtain a rate of order -1 the fast convergence behaviour of the large loss problem from Gordy and Juneja (2010) (cf. Proposition 1) must be, first, carried over to cdfs and then, second, to the quantile-based case at hand. The following result relies on a straightforward Taylor expansion of $F_{X+\epsilon_{N_2}}\left(q_\alpha^{X+\epsilon_{N_2}} \right)$. Indeed, the behaviour resp. the convergence of $F'_{X+\epsilon_{N_2}}(x)$ towards $F'_X(x)$, for $x \in \mathbb{R}$, is then one of the major challenges. This problem is, however, fully discussed in Boos (1985) (cf. Lemma 1) and Sweeting (1986) (cf. Theorem 1 and Theorem 2) under the heading of so-called local limit theorems. But, to obtain such a convergence result we rely, in the following, on the assumptions of Sweeting (1986) and thus introduce the *asymptotic uniform equicontinuity* condition:

Definition 4.2.4. *(cf. Sweeting (1986))*
A sequence $\left\{ F'_{X+\epsilon_{N_2}} \right\}$ *is asymptotically uniformly equicontinuous (a.u.e.c.) at* $x \in$
$\mathbb{R}$, *if for given* $\epsilon > 0$, *it exists a* $\delta(\epsilon)$ *and an* $\bar{N}_2(\epsilon)$ *such that whenever* $|y - x| \le \delta(\epsilon)$
holds, it applies

$$\forall N_2 \ge \bar{N}_2(\epsilon): \quad \left| F'_{X+\epsilon_{N_2}} (y) - F'_{X+\epsilon_{N_2}} (x) \right| < \epsilon.$$

Remark 4.2.5.
Note, that Assumption 1 implies immediately the previous a.u.e.c. condition due to a straightforward Taylor expansion. According to Assumption 1 (ii), as residual term the integral form can be used.

Based on these preliminary remarks and the definition of asymptotic uniform equicontinuity for densities we finally obtain:

Proposition 4.2.6.
Let Assumption 1 hold, then, for any $0 \le \beta < 1$, *there exists an* $\bar{N}_2 \in \mathbb{N}$ *such that*

$$\forall N_2 \ge \bar{N}_2: \quad \left| q_\alpha^{X+\epsilon_{N_2}} - q_\alpha^X \right| \le \frac{1}{2 \cdot N_2^\beta \cdot f_X \left(q_\alpha^X \right)}$$

holds.

Proof.
Since F_X and $F_{X+\epsilon_{N_2}}$ are cdfs, we immediately obtain

$$F_X \left(q_\alpha^X \right) - F_{X+\epsilon_{N_2}} \left(q_\alpha^X \right) = \mathbb{P} \left(X \le q_\alpha^X \right) - \mathbb{P} \left(X + \epsilon_{N_2} \le q_\alpha^X \right)$$
$$= \mathbb{P} \left(X + \epsilon_{N_2} > q_\alpha^X \right) - \mathbb{P} \left(X > q_\alpha^X \right).$$

This can be interpreted as a large loss problem (i.e. $G(X) = \mathbb{1}_{\{X > c\}}$) and Proposition 1 of Gordy and Juneja (2010) yields for $0 \le \beta < 1$ and under Assumption 1

$$F_X \left(q_\alpha^X \right) - F_{X+\epsilon_{N_2}} \left(q_\alpha^X \right) \le N_2^{-\beta}. \tag{4.1}$$

Applying a Taylor expansion at $q_\alpha^{X+\epsilon_{N_2}}$ yields for all N_2

$$F_{X+\epsilon_{N_2}}\left(q_\alpha^X\right) = F_{X+\epsilon_{N_2}}\left(q_\alpha^{X+\epsilon_{N_2}}\right) + F'_{X+\epsilon_{N_2}}(\xi) \cdot \left(q_\alpha^X - q_\alpha^{X+\epsilon_{N_2}}\right),$$

for $\xi \in \left[q_\alpha^X, q_\alpha^{X+\epsilon_{N_2}}\right]$. With (4.1), it holds

$$F_{X+\epsilon_{N_2}}\left(q_\alpha^X\right) \geq \alpha - N_2^{-\beta}$$

and thus, in concatenation with the Taylor expansion, we directly get

$$\alpha - N_2^{-\beta} \leq F_{X+\epsilon_{N_2}}\left(q_\alpha^{X+\epsilon_{N_2}}\right) + F'_{X+\epsilon_{N_2}}(\xi) \cdot \left(q_\alpha^X - q_\alpha^{X+\epsilon_{N_2}}\right)$$
$$= \alpha + F'_{X+\epsilon_{N_2}}(\xi) \cdot \left(q_\alpha^X - q_\alpha^{X+\epsilon_{N_2}}\right).$$

We know that $X + \epsilon_{N_2}$ converges in distribution to X. This, with the a.u.e.c. and boundedness assumption (cf. Assumption 1) on the series $\left\{F'_{X+\epsilon_{N_2}}\right\}$ yields $\|F'_{X+\epsilon_{N_2}} - F'_X\|_\infty = \sup_{x \in \mathbb{R}} |F'_{X+\epsilon_{N_2}}(x) - F'_X(x)| \to 0$ for $N_2 \to \infty$ (cf. Boos (1985), Lemma 1 resp. Sweeting (1986), Theorem 2). Hence, for continuous F'_X we obtain $F'_{X+\epsilon_{N_2}} \to F'_X$ as $N_2 \to \infty$ and for sufficiently large N_2 (i.e. $N_2 \geq \bar{N}_2$) $F'_{X+\epsilon_{N_2}}(\xi) \leq 2 \cdot F'_X\left(q_\alpha^X\right)$, i.e.

$$-N_2^{-\beta} \leq 2 \cdot F'_X\left(q_\alpha^X\right) \cdot \left(q_\alpha^X - q_\alpha^{X+\epsilon_{N_2}}\right)$$

and thus finally

$$q_\alpha^{X+\epsilon_{N_2}} - q_\alpha^X \leq N_2^{-\beta} \cdot \frac{1}{2 \cdot F'_X\left(q_\alpha^X\right)}.$$

Since the remaining inequality follows along similar lines, the claim is proven. $\square$

4.3   Main Results on a.s. Convergence

There are existing, similar as in the moment-based case, two different representations for the considered error term $\Delta\rho_{N_1,N_2} = \widehat{q}_{\alpha,N_1}^{X+\epsilon_{N_2}} - q_\alpha^X$, an *inner* as well as an *outer* representation. In the subsequent chapter, we rely on the classical *outer representation* given by

$$\Delta \rho_{N_1,N_2} = \widehat{q}_{\alpha,N_1}^{X+\epsilon_{N_2}} - q_\alpha^X = \widehat{q}_{\alpha,N_1}^{X+\epsilon_{N_2}} - q_\alpha^{X+\epsilon_{N_2}} + q_\alpha^{X+\epsilon_{N_2}} - q_\alpha^X$$

$$= \widehat{q}_{\alpha,N_1}^{X+\epsilon_{N_2}} - q_\alpha^{X+\epsilon_{N_2}} + Q_{N_2},$$

where $Q_{N_2} = q_\alpha^{X+\epsilon_{N_2}} - q_\alpha^X$ denotes the quantile difference between the distorted and true distribution function and can be – in some way – interpreted in this case as the bias. Remind, that we denote $\widehat{q}_{\alpha,N_1}^{X+\epsilon_{N_2}}$ as the quantile estimator of (1.11) based on the empirical distribution function $\widehat{F}_{X+\epsilon_{N_2}}^{N_1}$, i.e.

$$\widehat{q}_{\alpha,N_1}^{X+\epsilon_{N_2}} := \inf \left\{ x \in \mathbb{R} \mid \widehat{F}_{X+\epsilon_{N_2}}^{N_1}(x) \geq \alpha \right\},$$

for $0 < \alpha < 1$. Hence, we are now ready to formulate the main results of our contribution. Under very weak assumptions, especially on the budget sequence, Theorem 4.3.1 gives an exponential decay in the probability that the estimated quantile differs from the true quantile more than some ϵ. This yields, based on the complete convergence, to the desired convergence in the almost sure sense. Theorem 4.3.5 additionally provides a rate of the almost sure convergence if the budget sequence satisfies further restrictions.

Almost Sure Convergence without Convergence Rates

Theorem 4.3.1.
Let Assumptions (B0) *and* (B1,p) *hold for* $p \geq 2$. *Let further* $\{(N_1(N), N_2(N))\}$ *be an arbitrary budget sequence, then, for each* $t > 0$ *there exists an* $\bar{N} \in \mathbb{N}$ *such that*

$$\forall N \geq \bar{N}: \quad \mathbb{P}\left(\left| \widehat{q}_{\alpha,N_1}^{X+\epsilon_{N_2}} - q_\alpha^X \right| > t \right) \leq \bar{\lambda}^{N_1(N)}$$

holds, with $\bar{\lambda} = \exp\left(-\frac{\delta(t/12, X)^2}{2} \right)$. *Hence, the sequence* $\left\{ q_\alpha^{X+\epsilon_{N_2}} \right\}$ *converges in probability to* q_α^X. *If further* $\sum_{N=1}^\infty \lambda^{N_1(N)} < \infty$ *holds for all* $\lambda \in [0, 1[$, *then*

$$\widehat{q}_{\alpha,N_1}^{X+\epsilon_{N_2}} \xrightarrow[N\to\infty]{a.s.} q_\alpha^X.$$

Proof.
The first statement follows immediately from Lemma 4.3.4 and the fact that $\{(N_1(N), N_2(N))\}$ is a budget sequence according to Definition 2.1.1. Since $\lambda^{N_1(N)}$

goes to 0 for each $\lambda \in [0, 1[$, this applies also for the specific $\bar{\lambda}$. Hence, we immediately obtain the convergence in probability and the first statement. Almost sure convergence follows directly from complete convergence which holds since the sufficient condition $\sum_{N=1}^{\infty} \lambda^{N_1(N)} < \infty$ was assumed for all $\lambda \in [0, 1[$ and especially for $\bar{\lambda}$. $\qquad\square$

Note, that based on Hoeffdings inequality (cf. Corollary 2.3.2) we are, in this setting, able to derive an exponential decay in contrast to the polynomial decay in the previous moment-based setup. Furthermore, it should be highlighted that only very weak integrability assumptions on V and the locally non-flatness (B0) are needed to establish the stated convergence in probability for an arbitrary budget sequence. There is no need for strong smoothness assumptions as e.g. in Gordy and Juneja (2010). Moreover, the moment condition in (B1,p) is only required in Lemma 4.3.3 according to the usage of the Fuk-Nagaev inequality (cf. Corollary 2.2.2) and thus in this setup $p = 2$ is sufficient. Indeed, Lemma 4.3.3 can also be proven with the Markov inequality and thus with (B1, 1), but according to upcoming rate considerations the usage of the Fuk-Nagaev inequality is advantageous.

Remark 4.3.2.
For almost sure convergence a comparison to the convergent series $\sum_{N=1}^{\infty} N^{\eta}$ for $\eta < -1$ shows that it is sufficient to require

$$\frac{N_1(N)}{\ln(N)} \to \infty \quad for \ N \to \infty.$$

Hence, a minimum growth condition on $N_1(N)$ has to be fulfilled.

Next, the underlying technical results for Theorem 4.3.1 will be carried out. Here, we show first that the positivity of the slope carries over from the true distribution to the distorted distribution.

Lemma 4.3.3.
Let Assumptions (B0) and (B1,p) hold with $p \geq 2$. Then, for each $\epsilon > 0$ there exists an $\bar{N}_2 \in \mathbb{N}$ such that

$$\forall N_2 \geq \bar{N}_2 : \quad \delta(\epsilon, X + \epsilon_{N_2}) \geq \frac{1}{2} \delta\left(\frac{\epsilon}{4}, X\right) > 0.$$

Proof.
Since

$$\delta(\epsilon, X + \epsilon_{N_2}) = \min \left\{ \alpha - F_{X+\epsilon_{N_2}} \left(q_\alpha^{X+\epsilon_{N_2}} - \epsilon \right), F_{X+\epsilon_{N_2}} \left(q_\alpha^{X+\epsilon_{N_2}} + \epsilon \right) - \alpha \right\}$$

it is sufficient to show

$$F_{X+\epsilon_{N_2}} \left(q_\alpha^{X+\epsilon_{N_2}} + \epsilon \right) - \alpha > \frac{\delta \left(\frac{\epsilon}{4}, X \right)}{2}.$$

The remaining inequality follows analogous. Due to the monotonicity of distribution functions and (VQ, 1/4), we have for large enough N_2 due to Proposition 4.2.1

$$F_{X+\epsilon_{N_2}} \left(q_\alpha^{X+\epsilon_{N_2}} + \epsilon \right) \geq F_{X+\epsilon_{N_2}} \left(q_\alpha^X - \frac{1}{N_2^\beta} + \epsilon \right),$$

for some fixed $1/4 > \beta > 0$. By enlarging N_2 to satisfy $N_2 \geq \left\lceil \left(\frac{2}{\epsilon} \right)^{1/\beta} \right\rceil$, we continue by

$$F_{X+\epsilon_{N_2}} \left(q_\alpha^X - \frac{1}{N_2^\beta} + \epsilon \right) \geq F_{X+\epsilon_{N_2}} \left(q_\alpha^X + \frac{\epsilon}{2} \right).$$

We now rely again on Petrov (1975) (cf. Lemma 3) and obtain immediately

$$F_{X+\epsilon_{N_2}} \left(q_\alpha^X + \frac{\epsilon}{2} \right) \geq F_X \left(q_\alpha^X + \frac{\epsilon}{4} \right) - \mathbb{P} \left(|\epsilon_{N_2}| \geq \frac{\epsilon}{4} \right).$$

Thus, we get in summary

$$F_{X+\epsilon_{N_2}} \left(q_\alpha^{X+\epsilon_{N_2}} + \epsilon \right) - \alpha \geq \delta \left(\frac{\epsilon}{4}, X \right) - \mathbb{P} \left(|\epsilon_{N_2}| \geq \frac{\epsilon}{4} \right).$$

Since (B1,p) holds for $p \geq 2$, we obtain according to Corollary 2.2.2 with $t = \epsilon/4$ $\mathbb{P} \left(|\epsilon_{N_2}| \geq \frac{\epsilon}{4} \right) \to 0$ for $N_2 \to \infty$, and $\delta \left(\frac{\epsilon}{4}, X \right) > 0$ by (B0). Hence, the statement follows. $\qquad\square$

We now investigate the first term $\widehat{q}_{\alpha, N_1}^{X+\epsilon_{N_2}} - q_\alpha^{X+\epsilon_{N_2}}$ in the outer representation. In Lemma 4.3.4 we provide an exponentially decaying estimate (in N_1) for the probability that this difference is larger than some specified threshold if N_2 is large enough, i.e. if $q_\alpha^{X+\epsilon_{N_2}}$ is close enough to q_α^X. We can observe that this result holds without further assumptions on $F_{X+\epsilon_{N_2}}$, i.e. the sufficient condition for the almost sure convergence of the estimator for the true distribution F_X carries over to the distorted distribution $F_{X+\epsilon_{N_2}}$.

Lemma 4.3.4.
Let Assumption (B0) and Assumption (B1,p) hold, with $p \geq 2$. Then, for all $\epsilon > 0$ there exists some $\bar{N}_2 \in \mathbb{N}$ such that

$$\forall N_2 \geq \bar{N}_2, \forall N_1 \geq 1: \quad \mathbb{P}\left(\left|\widehat{q}_{\alpha,N_1}^{X+\epsilon_{N_2}} - q_\alpha^X\right| > \epsilon\right) \leq 2 \cdot \exp\left(-N_1 \cdot \frac{\delta\left(\frac{\epsilon}{12}, X\right)^2}{2}\right).$$

Proof.
Let $\tilde{\epsilon} = \epsilon/3$. With Proposition 4.2.1 we can choose N_2 sufficiently large such that $\left|q_\alpha^{X+\epsilon_{N_2}} - q_\alpha^X\right| \leq \tilde{\epsilon}$ holds. Hence, it is sufficient to show

$$\mathbb{P}\left(\left|\widehat{q}_{\alpha,N_1}^{X+\epsilon_{N_2}} - q_\alpha^{X+\epsilon_{N_2}}\right| > \tilde{\epsilon}\right) \leq 2 \cdot \exp\left(-N_1 \cdot \frac{\delta\left(\frac{\tilde{\epsilon}}{4}, X\right)^2}{2}\right).$$

In the following we use the distorted empirical distribution function $\widehat{F}_{X+\epsilon_{N_2}}^{N_1}$ and reformulate in accordance to Serfling (1980) (cf. Proof of Theorem 2.3.2 resp. Lemma 1.1.4) the original problem for $\widehat{q}_{\alpha,N_1}^{X+\epsilon_{N_2}}$ as follows:

$$\mathbb{P}\left(\widehat{q}_{\alpha,N_1}^{X+\epsilon_{N_2}} > q_\alpha^{X+\epsilon_{N_2}} + \tilde{\epsilon}\right) = \mathbb{P}\left(\alpha > \widehat{F}_{X+\epsilon_{N_2}}^{N_1}\left(q_\alpha^{X+\epsilon_{N_2}} + \tilde{\epsilon}\right)\right)$$

$$= \mathbb{P}\left(\alpha > \frac{1}{N_1}\sum_{i=1}^{N_1} \mathbb{1}_{\left\{X_i+\epsilon_{i,N_2} \leq q_\alpha^{X+\epsilon_{N_2}}+\tilde{\epsilon}\right\}}\right)$$

$$= \mathbb{P}\left(\alpha > \frac{1}{N_1}\sum_{i=1}^{N_1}\left(1 - \mathbb{1}_{\left\{X_i+\epsilon_{i,N_2} > q_\alpha^{X+\epsilon_{N_2}}+\tilde{\epsilon}\right\}}\right)\right)$$

Now, in order to simplify the notation we set $U_i := \mathbb{1}_{\left\{X_i+\epsilon_{i,N_2} > q_\alpha^{X+\epsilon_{N_2}}+\tilde{\epsilon}\right\}}$ and continue by

$$= \mathbb{P}\left(\sum_{i=1}^{N_1} U_i > N_1(1-\alpha)\right)$$

Since $\mathbb{E}[U_i] = \mathbb{E}\left[\mathbb{1}_{\left\{X_i+\epsilon_{i,N_2}>q_\alpha^{X+\epsilon_{N_2}}+\tilde{\epsilon}\right\}}\right] = 1 - F_{X+\epsilon_{N_2}}\left(q_\alpha^{X+\epsilon_{N_2}}+\tilde{\epsilon}\right)$ holds,

we obtain:

$$= \mathbb{P}\left(\sum_{i=1}^{N_1} U_i - \sum_{i=1}^{N_1} \mathbb{E}[U_i] > N_1\left(F_{X+\epsilon_{N_2}}(q_\alpha^{X+\epsilon_{N_2}}+\tilde{\epsilon}) - \alpha\right)\right)$$

$$\leq \mathbb{P}\left(\sum_{i=1}^{N_1} U_i - \sum_{i=1}^{N_1} \mathbb{E}[U_i] > N_1 \cdot \delta(\tilde{\epsilon}, X+\epsilon_{N_2})\right)$$

With Lemma 4.3.3 we immediately obtain for sufficiently large N_2:

$$\leq \mathbb{P}\left(\sum_{i=1}^{N_1} U_i - \sum_{i=1}^{N_1} \mathbb{E}[U_i] > N_1 \cdot \frac{1}{2} \cdot \delta\left(\frac{\tilde{\epsilon}}{4}, X\right)\right).$$

Based on these basic transformations we can now apply Hoeffding's Theorem (cf. Corollary 2.3.2) and obtain

$$\mathbb{P}\left(\widehat{q}_{\alpha,N_1}^{X+\epsilon_{N_2}} > q_\alpha^{X+\epsilon_{N_2}}+\tilde{\epsilon}\right) \leq \exp\left(-\frac{N_1 \cdot \delta\left(\frac{\tilde{\epsilon}}{4}, X\right)^2}{2}\right).$$

The remaining inequality follows along similar lines. Hence, the lemma is proven. $\qquad\square$

Almost Sure Convergence with Convergence Rates

In contrast to the previous almost sure considerations without rate factor we now need an additional assumption in our upcoming investigations, namely (VQ,β) for some $\beta > 0$. Similar as in the moment-based case with the vanishing bias rate, we need the vanishing quantile rate in order to obtain rates of convergence. Here, we investigate the (VQ,β) case for $0 \leq \beta < \min\left\{\frac{1}{2}, \frac{p-1}{p+1}\right\}$ under the weak Assumptions (B0) and (B1,p) for $p \geq 2$. Hence, in our rather general investigation it is sufficient that F_X is differentiable at q_α^X and that the tail of ϵ_{N_2} is sufficiently well behaved.

Theorem 4.3.5.

Let $p \geq 2$, $0 \leq \beta < \min\left\{\frac{1}{2}, \frac{p-1}{p+1}\right\}$ and suppose Assumptions (B0), (B1,p) and (VQ,β) hold. Further, let $0 < r < 1$ and consider the specific budget sequence $N_1(N) = \lfloor N^{1-r} \rfloor$, $N_2(N) = \lfloor N^r \rfloor$. Then, it holds:

(1.) If $0 < r < \frac{1}{1+2\beta}$ holds, then the sequence $N_2(N)^\beta \cdot \left|\widehat{q}_{\alpha,N_1(N)}^{X+\epsilon_{N_2(N)}} - q_\alpha^X\right|$ converges to 0 in probability. More exactly, for all $\epsilon > 0$ there exists an $\bar{N} \in \mathbb{N}$ such that:

$$\forall N \geq \bar{N}: \quad \mathbb{P}\left(N_2(N)^\beta \cdot \left|\widehat{q}_{\alpha,N_1(N)}^{X+\epsilon_{N_2(N)}} - q_\alpha^X\right| > \epsilon\right) \leq 2 \cdot \exp\left(-\frac{N_1(N) \cdot F_X'(q_\alpha^X)^2 \cdot \epsilon^2}{N_2(N)^{2\beta} \cdot 1152}\right)$$

(2.) If $0 < r < \frac{1}{1+2\beta}$, the sequence $N_2(N)^\beta \cdot \left|\widehat{q}_{\alpha,N_1(N)}^{X+\epsilon_{N_2(N)}} - q_\alpha^X\right|$ converges to 0 almost surely.

Proof.

The first statement follows immediately from Lemma 4.3.9 and the fact that $N_1(N) = \lfloor N^{1-r} \rfloor$, $N_2(N) = N^r$, for $0 < r < 1$. This is, according to Definition 2.1.1, a budget sequence. Since $\exp\left(-\frac{N_1(N) \cdot F_X'(q_\alpha^X)^2 \cdot \epsilon^2}{N_2(N)^{2\beta} \cdot 1152}\right)$ goes to 0 for each budget sequence with $N_1(N)/N_2(N)^{2\beta} \to \infty$, which holds for $0 < r < \frac{1}{1+2\beta}$, we obtain the desired convergence in probability.

The second statement and thus the almost sure convergence follows, again, directly from the complete convergence. This applies if the stricter condition $\sum_{N=1}^\infty \lambda^{N_1(N)/N_2(N)^{2\beta}} < \infty$ holds for $\lambda = \exp\left(-\frac{F_X'(q_\alpha^X)^2 \cdot \epsilon^2}{1152}\right)$. Here, a comparison to the convergent series $\sum_{N=1}^\infty N^{-\kappa}$ for $\kappa < -1$ yields that $\frac{N^{1-r-2\beta r}}{\ln(N)} \to \infty$ is sufficient for $N \to \infty$. This is in turn also fulfilled for $0 < r < \frac{1}{1+2\beta}$ since $N^{1-r-2\beta r}$ exceeds for large enough N the logarithm $\ln(N)$. $\square$

Remark 4.3.6.

Let us state here some helpful remarks resp. a brief overview over the obtainable rates:

- *The best convergence rate for $p \geq 3$ is then (almost) achieved if r is (almost) $1/2$ and β (almost) $1/2$ which gives a remarkably good convergence order of (almost) $-1/4$.*

- *For $p = 2$, the best choice for r is (almost) $3/5$ and β (almost) $1/3$ which gives a convergence order of (almost) $-1/5$.*
- *Compared to the RMSE convergence of order $-1/3$ in Gordy and Juneja (2010) this is still a quite remarkable result since our assumptions are significantly weaker compared to their strict smoothness and boundedness assumptions (cf. Assumption 1, Gordy and Juneja (2010)).*

For the proof of Theorem 4.3.5 we start completely analogous to the previous section with a corresponding lemma, quantifying the size of the underlying slope by considering some additional rate factor N_2^β:

Lemma 4.3.7.
Let Assumptions (B0) and (B1,p) hold with $p \geq 2$. Suppose further that (VQ,β) holds with $0 \leq \beta < \min\left\{\frac{1}{2}, \frac{p-1}{p+1}\right\}$. Then, for each $\epsilon > 0$ there exists an $\bar{N}_2 \in \mathbb{N}$ such that

$$\forall N_2 \geq \bar{N}_2: \quad \delta\left(\frac{\epsilon}{N_2^\beta}, X + \epsilon_{N_2}\right) \geq F_X'(q_\alpha^X) \cdot \frac{\epsilon}{16N_2^\beta} > 0.$$

Proof.
Similar as in the proof of Lemma 4.3.3, it is sufficient to show

$$F_{X+\epsilon_{N_2}}\left(q_\alpha^{X+\epsilon_{N_2}} + \frac{\epsilon}{N_2^\beta}\right) - \alpha > F_X'(q_\alpha^X) \cdot \frac{\epsilon}{16N_2^\beta}.$$

We use the monotonicity of the distribution function and (VQ,β) (with $\epsilon/2$) and obtain for large enough N_2

$$F_{X+\epsilon_{N_2}}\left(q_\alpha^{X+\epsilon_{N_2}} + \frac{\epsilon}{N_2^\beta}\right) \geq F_{X+\epsilon_{N_2}}\left(q_\alpha^X + \frac{\epsilon}{2N_2^\beta}\right)$$

Using Lemma 3 of Petrov (1975), once again, leads directly to

$$F_{X+\epsilon_{N_2}}\left(q_\alpha^X + \frac{\epsilon}{2N_2^\beta}\right) \geq F_X\left(q_\alpha^X + \frac{\epsilon}{4N_2^\beta}\right) - \mathbb{P}\left(|\epsilon_{N_2}| \geq \frac{\epsilon}{4N_2^\beta}\right).$$

Similarly to the proof of Proposition 4.2.1, we get, due to the differentiability assumption (B0), for sufficiently large N_2:

$$F_{X+\epsilon_{N_2}}\left(q_\alpha^X + \frac{\epsilon}{2N_2^\beta}\right) \geq \alpha + F_X'(q_\alpha^X) \cdot \frac{\epsilon}{8N_2^\beta} - \mathbb{P}\left(|\epsilon_{N_2}| \geq \frac{\epsilon}{4N_2^\beta}\right).$$

Now, according to Corollary 2.2.3, the term $\mathbb{P}\left(|\epsilon_{N_2}| \geq \frac{\epsilon}{4N_2^\beta}\right)$ can be bounded by $D_p \cdot N_2^{-p(1-\beta)+1}$, whereby D_p is a constant depending on the number of existing moments in (B1,p). Indeed, the statement follows now for sufficiently large N_2 since $\beta < \frac{p-1}{p+1}$. $\qquad\square$

Remark 4.3.8.

Here, some remarks seem to be adequate. Importantly, it should be noted that the previous Lemma 4.3.7 relies on Lemma 3 of Petrov (1975) and in the following on the Fuk-Nagaev inequality (cf. Corollary 2.2.3). Since the Fuk-Nagaev inequality quantifies the speed of convergence in the LLN, it allows a maximum order of (almost) $1/2$ for $p > 2$. Hence, all (VQ,β) rates for $\beta \geq 1/2$ deliver in this proof strategy no further advantages since they will be truncated by the Fuk-Nagaev inequality. If a faster (VQ,β) rate than $\beta \geq 1/2$ will be applied, we must also rely on strict smoothness assumptions as mentioned in Assumption 1 resp. Gordy and Juneja (2010). Because then by a Taylor expansion and the boundedness of the partial derivatives, it is possible to obtain a $\delta\left(\epsilon, X + \epsilon_{N_2}\right)$ approximation which exploits the fast (VQ,β) rate with $\beta \geq 1/2$.

Lemma 4.3.9.

Let Assumptions (B0) and (B1,p) hold with $p \geq 2$. Suppose further that (VQ,β) applies for $0 \leq \beta < \min\left\{\frac{1}{2}, \frac{p-1}{p+1}\right\}$. Then, for all $\epsilon > 0$ there exists some $\bar{N}_2 \in \mathbb{N}$ such that

$$\forall N_2 \geq \bar{N}_2 : \quad \mathbb{P}\left(\left|\widehat{q}_{\alpha,N_1}^{X+\epsilon_{N_2}} - q_\alpha^X\right| > \frac{\epsilon}{N_2^\beta}\right) \leq 2 \cdot \exp\left(-\frac{N_1 \cdot F_X'(q_\alpha^X)^2 \cdot \epsilon^2}{N_2^{2\beta} \cdot 1152}\right).$$

Proof.

Let $\tilde{\epsilon} = \frac{\epsilon}{3}$. By Assumption (VQ,β), we can choose N_2 sufficiently large such that $\left|q_\alpha^{X+\epsilon_{N_2}} - q_\alpha^X\right| \leq \frac{\tilde{\epsilon}}{N_2^\beta}$ holds. Hence, it is sufficient to prove

$$\mathbb{P}\left(\left|\widehat{q}_{\alpha,N_1}^{X+\epsilon_{N_2}} - q_\alpha^{X+\epsilon_{N_2}}\right| > \frac{\tilde{\epsilon}}{N_2^\beta}\right) \leq 2 \cdot \exp\left(-\frac{N_1 \cdot F_X'(q_\alpha^X)^2 \cdot \tilde{\epsilon}^2}{N_2^{2\beta} \cdot 128}\right).$$

Once again we use the distorted empirical distribution function $\widehat{F}_{X+\epsilon_{N_2}}^{N_1}$ and reformulate in accordance to Serfling (1980) (cf. Proof of Theorem 2.3.2 resp. Lemma 1.1.4) the original problem for $\widehat{q}_{\alpha,N_1}^{X+\epsilon_{N_2}}$ as follows:

$$\mathbb{P}\left(\widehat{q}_{\alpha,N_1}^{X+\epsilon_{N_2}} > q_\alpha^{X+\epsilon_{N_2}} + \frac{\tilde{\epsilon}}{N_2^\beta}\right) = \mathbb{P}\left(\alpha > \widehat{F}_{X+\epsilon_{N_2}}^{N_1}\left(q_\alpha^{X+\epsilon_{N_2}} + \frac{\tilde{\epsilon}}{N_2^\beta}\right)\right)$$

$$= \mathbb{P}\left(\alpha > \frac{1}{N_1}\sum_{i=1}^{N_1}\mathbb{1}_{\left\{X_i+\epsilon_{i,N_2}\leq q_\alpha^{X+\epsilon_{N_2}} + \frac{\tilde{\epsilon}}{N_2^\beta}\right\}}\right)$$

$$= \mathbb{P}\left(\alpha > \frac{1}{N_1}\sum_{i=1}^{N_1}\left(1-\mathbb{1}_{\left\{X_i+\epsilon_{i,N_2}> q_\alpha^{X+\epsilon_{N_2}} + \frac{\tilde{\epsilon}}{N_2^\beta}\right\}}\right)\right)$$

In order to simplify the notation, we set $U_i := \mathbb{1}_{\left\{X_i+\epsilon_{i,N_2}> q_\alpha^{X+\epsilon_{N_2}} + \frac{\tilde{\epsilon}}{N_2^\beta}\right\}}$ and continue by

$$= \mathbb{P}\left(\sum_{i=1}^{N_1}U_i > N_1(1-\alpha)\right).$$

Since $\mathbb{E}[U_i] = \mathbb{E}\left[\mathbb{1}_{\left\{X_i+\epsilon_{i,N_2}> q_\alpha^{X+\epsilon_{N_2}} + \frac{\tilde{\epsilon}}{N_2^\beta}\right\}}\right] = 1 - F_{X+\epsilon_{N_2}}\left(q_\alpha^{X+\epsilon_{N_2}} + \frac{\tilde{\epsilon}}{N_2^\beta}\right)$ holds, we obtain:

$$= \mathbb{P}\left(\sum_{i=1}^{N_1}U_i - \sum_{i=1}^{N_1}\mathbb{E}[U_i] > N_1\left(F_{X+\epsilon_{N_2}}\left(q_\alpha^{X+\epsilon_{N_2}} + \frac{\tilde{\epsilon}}{N_2^\beta}\right) - \alpha\right)\right)$$

$$\leq \mathbb{P}\left(\sum_{i=1}^{N_1}U_i - \sum_{i=1}^{N_1}\mathbb{E}[U_i] > N_1\cdot\delta\left(\frac{\tilde{\epsilon}}{N_2^\beta}, X+\epsilon_{N_2}\right)\right)$$

According to Lemma 4.3.7, we obtain a sufficiently large N_2 such that

$$\leq \mathbb{P}\left(\sum_{i=1}^{N_1}U_i - \sum_{i=1}^{N_1}\mathbb{E}[U_i] > N_1\cdot F_X'(q_\alpha^X)\cdot\frac{\tilde{\epsilon}}{16\cdot N_2^\beta}\right).$$

Finally, we again apply Hoeffding's Theorem (cf. Corollary 2.3.2) and obtain

$$\mathbb{P}\left(\widehat{q}_{\alpha,N_1}^{X+\epsilon_{N_2}} > q_\alpha^{X+\epsilon_{N_2}} + \frac{\tilde{\epsilon}}{N_2^\beta}\right) \leq \exp\left(-\frac{N_1 \cdot F_X'(q_\alpha^X)^2 \cdot \tilde{\epsilon}^2}{N_2^{2\beta} \cdot 128}\right).$$

The remaining inequality follows along similar lines and thus the claim is proven.   □

Remark 4.3.10.

It should be highlighted that the choice of $\bar{N}_2$ in Lemma 4.3.9 is independent of the number of outers, i.e. N_1. Hence, Lemma 4.3.9 applies for all $N_1 \in \mathbb{N}$.

Non Parametric Confidence Intervals for Quantiles

This Chapter is intended to introduce a new approach in order to obtain model free and in this nested Monte Carlo setup noise considering confidence intervals for the quantile q_α^X. As a brief reminder X evolves here from the nested simulation. Therefore, we particularly extend the non-noise considering approach of David and Nagaraja (2004) to the problem at hand. This whole Chapter grounds on the results of Klein and Werner (2023a) and we divide the upcoming chapter into the following sections:

According to the extra inner simulation an additional noise level, i.e. ϵ_{N_2}, occurs in nested simulations compared to standard non nested simulations. Hence, based on a brief example, we first motivate noise considering confidence intervals in Section 5.1 by pointing out that non-noise considering confidence intervals are not sufficient according to the occurring noise level. This problem happens especially for small N_2 and thus in noisy environments. Then, in Section 5.2 we summarize the needed weak and few assumptions to obtain confidence intervals and on top also a corresponding error estimate. The aim is to introduce a rather general setup with as few assumptions on the cdf F_X as possible. Thus, quantile functions will be repeated and addressed in general. Finally, Section 5.3 provides our novel result in Theorem 5.3.3 which states the noise considering asymptotic non parametric confidence interval method for quantiles q_α^X. According to the model free approach, we rely on the sample variance since σ^2 is generally unknown. Hence, our additional error estimate is based on a customized version of the well-known Berry-Esseen Theorem (cf. Theorem 2.4.2). Moreover, it should be highlighted that we are able to simplify and improve the related but also rather complex approach of Lan et al. (2007b). Our numerical simulations, which are summarized in Chapter 6, underpin

© The Author(s), under exclusive license to Springer Fachmedien Wiesbaden GmbH, part of Springer Nature 2024
M. Klein, *Nested Simulations: Theory and Application*, Mathematische Optimierung und Wirtschaftsmathematik | Mathematical Optimization and Economathematics, https://doi.org/10.1007/978-3-658-43853-1_5

this last statement. Here, we rely, first, on simulations based on an academic test example (cf. Section 6.1) and second, to show its practicability on simulations based on the insurance model of Hieber et al. (2019) (see Section 6.2).

5.1   Motivation

Assuming the standard case that X follows a continuous distribution function and each $X_1, \ldots, X_{N_1}$ is known explicitly, i.e. each outer node X_i contains the true conditional expectation $X_i = \mathbb{E}[V_i | Z = Z_i]$, the derivation of confidence intervals for quantiles is straightforward see e.g. Hult et al. (2012). As shortly mentioned in the introduction, David and Nagaraja (2004) introduce a more general framework by allowing also discrete distribution functions for X. To find a confidence interval for the α−quantile which fulfills for given $0 < \epsilon < 1$

$$\mathbb{P}\left(q_\alpha^X \in [LB, UB]\right) \geq 1 - \epsilon,$$

we consider, according to David and Nagaraja (2004),

$$
\begin{aligned}
\mathbb{P}\left(X_{(r_\epsilon)} \leq q_\alpha^X \leq X_{(s_\epsilon)}\right) &\geq \mathbb{P}\left(X_{(r_\epsilon)} \leq q_\alpha^X\right) - \mathbb{P}\left(X_{(s_\epsilon)} < q_\alpha^X\right) \\
&= \mathbb{P}\left(\#\left\{i | X_i \leq q_\alpha^X\right\} \geq r_\epsilon\right) - \mathbb{P}\left(\#\left\{i | X_i < q_\alpha^X\right\} \geq s_\epsilon\right) \\
&= \mathbb{P}\left(\#\left\{i | X_i > q_\alpha^X\right\} \leq N_1 - r_\epsilon\right) - \mathbb{P}\left(\#\left\{i | X_i \geq q_\alpha^X\right\} \leq N_1 - s_\epsilon\right) \\
&= \mathbb{P}\left(\sum_{i=1}^{N_1} \mathbb{1}_{\{X_i > q_\alpha^X\}} \leq N_1 - r_\epsilon\right) - \mathbb{P}\left(\sum_{i=1}^{N_1} \mathbb{1}_{\{X_i \geq q_\alpha^X\}} \leq N_1 - s_\epsilon\right) \\
&= \mathbb{P}\left(\sum_{i=1}^{N_1} \mathbb{1}_{\left\{X_i \leq q_\alpha^X\right\}} \geq r_\epsilon\right) - \mathbb{P}\left(\sum_{i=1}^{N_1} \mathbb{1}_{\left\{X_i < q_\alpha^X\right\}} \geq s_\epsilon\right) \\
&= 1 - \mathbb{P}\left(\sum_{i=1}^{N_1} \mathbb{1}_{\left\{X_i \leq q_\alpha^X\right\}} < r_\epsilon\right) - \left[1 - \mathbb{P}\left(\sum_{i=1}^{N_1} \mathbb{1}_{\left\{X_i < q_\alpha^X\right\}} < s_\epsilon\right)\right] \\
&= \mathbb{P}\left(\sum_{i=1}^{N_1} \mathbb{1}_{\left\{X_i < q_\alpha^X\right\}} < s_\epsilon\right) - \mathbb{P}\left(\sum_{i=1}^{N_1} \mathbb{1}_{\left\{X_i \leq q_\alpha^X\right\}} < r_\epsilon\right).
\end{aligned}
$$

Since $\sum_{i=1}^{N_1} \mathbb{1}_{\{X_i \leq q_\alpha^X\}}$ and $\sum_{i=1}^{N_1} \mathbb{1}_{\{X_i < q_\alpha^X\}}$ are binomial distributed with parameter N_1, probabilities $\mathbb{P}\left(X \leq q_\alpha^X\right) \geq \alpha$ (and $\alpha \geq \mathbb{P}\left(X < q_\alpha^X\right)$) and, according to the fact that the binomial distribution is monotonically decreasing in p, we obtain

$$
\geq \sum_{j=0}^{s_\epsilon-1} \binom{N_1}{j} \alpha^j (1-\alpha)^{N_1-j} - \sum_{j=0}^{r_\epsilon-1} \binom{N_1}{j} \alpha^j (1-\alpha)^{N_1-j}
$$

$$
= \sum_{j=r_\epsilon}^{s_\epsilon-1} \binom{N_1}{j} \alpha^j (1-\alpha)^{N_1-j},
$$

whereby $X_{(l)}$ denotes the l-th order statistic from $\{X_i\}_{i=1}^{N_1}$. Overall, for the determination of a $(1-\epsilon)$-confidence interval it is thus sufficient to seek an r_ϵ and s_ϵ such that $\sum_{j=r_\epsilon}^{s_\epsilon-1} \binom{N_1}{j} \alpha^j (1-\alpha)^{N_1-j} \geq 1-\epsilon$ is fulfilled and $LB = X_{(r_\epsilon)}$ and $UB = X_{(s_\epsilon)}$ can be set.

The exact knowledge of X_i is a simplification of the nested simulation approach at hand but illustrates some of the methods that we will enhance in the course of this work. The shortcomings of confidence interval methods without noise consideration are illustrated in the following brief example where we rely on the simulation framework from Example 6.1.2, i.e. $Z \sim \mathcal{N}(0,1)$ and $V|Z \sim \mathcal{LN}(0, Z^2/t)$. Here, for $t = 10$ and $\alpha = 99.5\%$ we obtain $-q_\alpha^X \approx -1.482859$ (cf. Appendix A.2). Due to our insurance specific interest we consider the negative quantile since $-q_\alpha^X = VaR_\alpha(X)$ holds and the VaR is the SCR relevant risk measure. Since for n identical simulation runs a $(1-\epsilon)$-confidence interval with LB and UB should capture the unknown q_α^X in approximately at least $(1-\epsilon) \cdot n$ cases, this leads to the conclusion that approximately $\epsilon \cdot n$ misfits, i.e. $q_\alpha^X \notin [LB, UB]$, should occur. Thus, based on 1000 simulation runs we would expect a maximal amount of at most 10 misfits for a given confidence level of $\epsilon = 0.01$. But Figure 5.1 indicates clearly that the commonly used confidence interval method without noise consideration clearly fails in the underlying noisy environment, i.e. in a range from 2 up to roughly 10^2 inners, according to the high noise level in the simulation. Nevertheless, for $N_2 > 10^2$ the noise level seems to be neglectable and thus, also the standard confidence interval method delivers quite reasonable intervals. This motivating example shows clearly that adjusted confidence interval methods for nested Monte Carlo simulations are necessary to capture the occurring noise level from the inner simulation. This is especially needed for simulation environments with few inner scenarios, a very common case in practice since inner simulations are expensive. Thus, in the

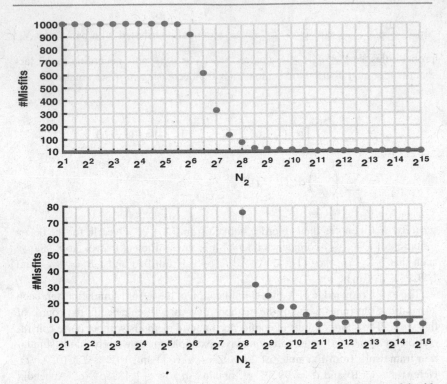

Figure 5.1 Numerical simulation study of an analytical nested Monte Carlo model with $N_1 = 10^4$ fixed and rising $N_2 = 2^1, \ldots, 2^{15}$. Plot of the occurring number of misfits according to the non-noise considering interval approach (cf. David and Nagaraja (2004)) based on 1000 simulation runs. Note, that for $N_2 = 1$ we obtain $\sigma(X) = 0.0793$ and $\sigma(\epsilon_{N_2}) = 0.3783$ as standard deviations

following we focus on the derivation of model free asymptotic confidence intervals which take the occurring noise level into account and satisfy for fixed $\epsilon > 0$ and $\alpha \in (0, 1)$

$$\forall N_1 \geq \bar{N} : \quad \lim_{N_2 \to \infty} \mathbb{P}\left(\widehat{LB}_{N_2,\epsilon} \leq q_\alpha^X \leq \widehat{UB}_{N_2,\epsilon}\right) \geq 1 - \epsilon, \qquad (5.1)$$

asymptotically in N_2. Note, that in this non parametric setup no additional assumptions on the variance of $V|Z$ or $f_X(q_\alpha^X)$ are needed. Compared to the already mentioned convergence in distribution results of Lee (1998) and Liu et al. (2022), this is a remarkable benefit since there the variance σ^2 of $V|Z$ is needed. Lan et al. (2007b) instead relies on similar assumptions like our approach and obtains also asymptotic confidence intervals for fixed N_1 (cf. Section 1.4), albeit without proof. But we will improve this approach further for the VaR and provide additionally quantitative error estimates.

5.2 Assumptions and Definitions

This section is intended to summarize briefly the necessary assumptions and to define quantile functions in general.

Assumptions

For our following considerations we need one rather weak integrability assumption and a non-flatness assumption on F_X. In order to apply a special version of the Berry-Esseen Theorem and to obtain thus a quantitative error estimate, V needs at least three moments, i.e.

$$V \in L^3(\Omega, \mathcal{F}, \mathbb{P}), \tag{C}$$

whereby this implies immediately $X \in L^3(\Omega, \mathcal{F}, \mathbb{P})$ and $\epsilon_{N_2} \in L^3(\Omega, \mathcal{F}, \mathbb{P})$ for each N_2, cf. Proposition 1.4.1. Similar to Chapter 4, we make the following non-flatness assumption, such that

$$\delta(\epsilon, X) := \min\left\{\alpha - F_X(q_\alpha^X - \epsilon), F_X(q_\alpha^X + \epsilon) - \alpha\right\} \tag{NF}$$

is strictly positive for all $\epsilon > 0$.

Quantile Functions

To obtain the most general setup in the following analysis also discrete distribution functions should be allowed. This has severe consequence on the quantile and thus we provide a short recap of general *quantile functions* and their properties here.

Definition 5.2.1.
An inverse function $q^X : (0, 1) \to \mathbb{R}$ of a cumulative distribution function F_X, of a random variable X, is called quantile function, if for all $\alpha \in (0, 1)$

$$\lim_{\eta \uparrow \alpha} F_X \left(q_\eta^X \right) \leq \alpha \leq \lim_{\eta \downarrow \alpha} F_X \left(q_\eta^X \right)$$

holds. Thereby, we denote for later considerations

$$q_{\alpha,-}^X = \sup \{x \in \mathbb{R} \mid \mathbb{P}(X \leq x) < \alpha\} \quad and \quad q_{\alpha,+}^X = \inf \{x \in \mathbb{R} \mid \mathbb{P}(X \leq x) > \alpha\},$$

as the lower and upper quantile functions of F_X (cf. Föllmer and Schied (2016)).

Remark 5.2.2.
Based on that definition two immediate and important consequences can immediately be made:

- *According to Lemma A.19. in Föllmer and Schied (2016) $q_{\alpha,-}^X \leq q_\alpha^X \leq q_{\alpha,+}^X$ holds generally for all $\alpha \in (0, 1)$, i.e. in particular for all underlying distribution functions F_X.*
- *If, on the other hand, the distribution function F_X is non-flat (cf. (NF)), as often the case in insurance specific setups, we immediately obtain $q_{\alpha,-}^X = q_\alpha^X = q_{\alpha,+}^X$ for all $\alpha \in (0, 1)$.*

5.3   Main Results on Confidence Intervals

In this section we are focusing on the theory of asymptotic confidence intervals for nested simulations. We proof and demonstrate a new approach in constructing model free asymptotic confidence intervals for quantiles and thus also the Value-at-Risk. In summary, we first build—based on the CLT and similar to Lan et al. (2007b)—asymptotic confidence intervals around each observation X_i, in order to obtain lower and upper bounds (i.e. $\bar{X}_i^l \leq X_i \leq \bar{X}_i^u$) such that $X_i \in [\bar{X}_i^l, \bar{X}_i^u]$ holds with a coverage probability of at least $1 - \delta$ for $\delta \in (0, 1)$. Then, second, we enhance the non-noise considering confidence interval approach of David and Nagaraja (2004) by considering two order statistics instead of one. Particularly, we take the order statistic of $\{\bar{X}_i^l\}$ and $\{\bar{X}_i^u\}$ for $i = 1, \ldots, N_1$. First, let us construct on each X_i an $(1 - \delta)$-asymptotic confidence interval.

Lemma 5.3.1.
If $V \in L^2(\Omega, \mathcal{F}, \mathbb{P})$, then for given CLT level $\delta \in (0, 1)$ it holds that

$$\forall i = 1, \ldots, N_1 : \quad \lim_{N_2 \to \infty} \mathbb{P} \left(\bar{X}_i - \Delta_{i,N_2} \leq X_i \leq \bar{X}_i + \Delta_{i,N_2} \right) \geq 1 - \delta,$$

whereby $\Delta_{i,N_2} = \Phi^{-1}_{1-\frac{\delta}{2}} \cdot \frac{\hat{\sigma}_i}{\sqrt{N_2}}$ *and* $\hat{\sigma}_i := \frac{1}{N_2} \sum_{j=1}^{N_2} (V_j - \bar{X}_i)$ *denotes the sample standard deviation.*

Proof.
Based on basic transformations and an application of Theorem 2.4.1 (CLT for empirical standard deviation) the claim follows immediately. $\square$

Note, that this approach is intended to take the arising error of the inner simulation into account. Since the variance of each X_i is unknown in this framework, the CLT based on the sample standard deviation $\hat{\sigma}_i$ was used. Consequently, after building confidence intervals around every outer node, in a second step, the CLT probabilities and the quantile function must be connected. This gap is closed in the subsequent proposition.

Proposition 5.3.2.
Let $V \in L^3(\Omega, \mathcal{F}, \mathbb{P})$, *then for given quantile level* $\alpha \in (0, 1)$ *and CLT level* $\delta \in (0, 1)$ *it holds*

$$\forall i = 1, \ldots, N_1 : \quad \mathbb{P}\left(\bar{X}_i + \Delta_{i,N_2} < q^X_{\alpha,-}\right) \le \alpha + \frac{\delta}{2} \pm \mathcal{O}\left(N_2^{-1/2}\right)$$

as well as

$$\forall i = 1, \ldots, N_1 : \quad \mathbb{P}\left(\bar{X}_i - \Delta_{i,N_2} \le q^X_{\alpha,+}\right) \ge \alpha - \frac{\delta}{2} \pm \mathcal{O}\left(N_2^{-1/2}\right).$$

Proof.
We start by noting that $\bar{X}_i = X_i + \epsilon_{i,N_2}$. Hence, it holds

$$\mathbb{P}\left(\bar{X}_i + \Delta_{i,N_2} < q^X_{\alpha,-}\right) = \mathbb{P}\left(X_i + \epsilon_{i,N_2} < q^X_{\alpha,-} - \Delta_{i,N_2}\right).$$

Then, Lemma 3 of Petrov (1975) yields for $\epsilon = \Delta_{i,N_2} > 0$

$$\mathbb{P}\left(X_i + \epsilon_{i,N_2} < q^X_{\alpha,-} - \Delta_{i,N_2}\right) \le F_X\left(q^X_{\alpha,-}\right) + \mathbb{P}\left(|\epsilon_{i,N_2}| \ge \Delta_{i,N_2}\right).$$

This in combination with Definition 5.2.1 leads straightforwardly to

$$\mathbb{P}\left(X_i + \epsilon_{i,N_2} \leq q^X_{\alpha,-} - \Delta_{i,N_2}\right) \leq \alpha + \mathbb{P}\left(|\epsilon_{i,N_2}| \geq \Delta_{i,N_2}\right)$$
$$= \alpha + 1 - \mathbb{P}\left(|\epsilon_{i,N_2}| < \Delta_{i,N_2}\right)$$
$$= \alpha + 1 - \mathbb{P}\left(\frac{\sqrt{N_2} \cdot |\epsilon_{i,N_2}|}{\hat{\sigma}_i} < \Phi^{-1}\left(1 - \frac{\delta}{2}\right)\right)$$

Next, we want to apply the Berry-Esseen Theorem (cf. Theorem 2.4.2), but note that here the sample standard deviation is involved since the analytical standard deviation is generally unknown. Thus, we rely instead on Theorem 1.1 of Bentkus and Götze (1996) and obtain for sufficiently large N_2:

$$\leq \alpha + 1 - \Phi\left(\Phi^{-1}\left(1 - \frac{\delta}{2}\right)\right) \pm \mathcal{O}\left(N_2^{-1/2}\right)$$
$$= \alpha + \frac{\delta}{2} \pm \mathcal{O}\left(N_2^{-1/2}\right).$$

Overall, this yields

$$\mathbb{P}\left(\bar{X}_i + \Delta_{i,N_2} \leq q^X_{\alpha,-}\right) \leq \alpha + \frac{\delta}{2} \pm \mathcal{O}\left(N_2^{-1/2}\right).$$

Since the remaining inequality follows along similar lines, the claim is proven here. $\qquad\square$

We are now able to derive the following asymptotic confidence interval for the quantile q^X_α. It should—once again—be highlighted that this result enables a model free approach, i.e. without cdf assumptions, since it is possible to rely on the sample variance.

Theorem 5.3.3.
Suppose $V \in L^3(\Omega, \mathcal{F}, \mathbb{P})$ and (NF) hold. Then, for any $\epsilon > 0$, quantile level $0 < \alpha < 1$ and CLT level $\delta > 0$ there exists an $\bar{N}_1 := \bar{N}_1(\epsilon, \alpha, \delta) \in \mathbb{N}$ such that

$$\forall N_1 \geq \bar{N}_1: \quad \mathbb{P}\left(\widehat{LB}_{N_2,\epsilon} \leq q^X_\alpha \leq \widehat{UB}_{N_2,\epsilon}\right) \geq 1 - \epsilon \pm \mathcal{O}\left(N_2^{-1/2}\right)$$

holds. Note, that $\left(\bar{X} - \Delta_{N_2}\right)_{(n)}$ and $\left(\bar{X} + \Delta_{N_2}\right)_{(n)}$ denote the n-th order statistic of $\left\{\bar{X}_i - \Delta_{i,N_2}\right\}_{i=1}^{N_1}$ and $\left\{\bar{X}_i + \Delta_{i,N_2}\right\}_{i=1}^{N_1}$. Hence, at least one r_ϵ and s_ϵ can be found such that $\widehat{LB}_{N_2,\epsilon} = \left(\bar{X} - \Delta_{N_2}\right)_{(r_\epsilon)}$ serves as lower and $\widehat{UB}_{N_2,\epsilon} = \left(\bar{X} + \Delta_{N_2}\right)_{(s_\epsilon)}$ as upper bound.

Proof.
Similar to the motivation we split the probability and obtain

$$
\mathbb{P}\left(\left(\bar{X} - \Delta_{N_2}\right)_{(r_\epsilon)} \leq q_\alpha^X \leq \left(\bar{X} + \Delta_{N_2}\right)_{(s_\epsilon)}\right)
$$
$$
\geq \mathbb{P}\left(\left(\bar{X} - \Delta_{N_2}\right)_{(r_\epsilon)} \leq q_\alpha^X\right) - \mathbb{P}\left(\left(\bar{X} + \Delta_{N_2}\right)_{(s_\epsilon)} < q_\alpha^X\right)
$$
$$
= \mathbb{P}\left(\#\left\{i \,|\, \bar{X}_i - \Delta_{i,N_2} \leq q_\alpha^X\right\} \geq r_\epsilon\right) - \mathbb{P}\left(\#\left\{i \,|\, \bar{X}_i + \Delta_{i,N_2} < q_\alpha^X\right\} \geq s_\epsilon\right).
$$

Now, the complement rule yields

$$
= \mathbb{P}\left(N_1 - \#\left\{i \,|\, \bar{X}_i - \Delta_{i,N_2} > q_\alpha^X\right\} \geq r_\epsilon\right) - \mathbb{P}\left(N_1 - \#\left\{i \,|\, \bar{X}_i + \Delta_{i,N_2} \geq q_\alpha^X\right\} \geq s_\epsilon\right)
$$
$$
= \mathbb{P}\left(\#\left\{i \,|\, \bar{X}_i - \Delta_{i,N_2} > q_\alpha^X\right\} \leq N_1 - r_\epsilon\right) - \mathbb{P}\left(\#\left\{i \,|\, \bar{X}_i + \Delta_{i,N_2} \geq q_\alpha^X\right\} \leq N_1 - s_\epsilon\right).
$$

The number operator can be expressed as a sum of indicator functions over the corresponding subset, i.e.

$$
\mathbb{P}\left(\#\left\{i \,|\, \bar{X}_i - \Delta_{i,N_2} > q_\alpha^X\right\} \leq N_1 - r_\epsilon\right) - \mathbb{P}\left(\#\left\{i \,|\, \bar{X}_i + \Delta_{i,N_2} \geq q_\alpha^X\right\} \leq N_1 - s_\epsilon\right)
$$
$$
= \mathbb{P}\left(\sum_{i=1}^{N_1} \mathbb{1}_{\{\bar{X}_i - \Delta_{i,N_2} > q_\alpha^X\}} \leq N_1 - r_\epsilon\right) - \mathbb{P}\left(\sum_{i=1}^{N_1} \mathbb{1}_{\{\bar{X}_i + \Delta_{i,N_2} \geq q_\alpha^X\}} \leq N_1 - s_\epsilon\right)
$$
$$
= \mathbb{P}\left(\sum_{i=1}^{N_1} \mathbb{1}_{\{\bar{X}_i - \Delta_{i,N_2} \leq q_\alpha^X\}} \geq r_\epsilon\right) - \mathbb{P}\left(\sum_{i=1}^{N_1} \mathbb{1}_{\{\bar{X}_i + \Delta_{i,N_2} < q_\alpha^X\}} \geq s_\epsilon\right)
$$
$$
= 1 - \mathbb{P}\left(\sum_{i=1}^{N_1} \mathbb{1}_{\{\bar{X}_i - \Delta_{i,N_2} \leq q_\alpha^X\}} < r_\epsilon\right) - \left[1 - \mathbb{P}\left(\sum_{i=1}^{N_1} \mathbb{1}_{\{\bar{X}_i + \Delta_{i,N_2} < q_\alpha^X\}} < s_\epsilon\right)\right]
$$
$$
= \mathbb{P}\left(\sum_{i=1}^{N_1} \mathbb{1}_{\{\bar{X}_i + \Delta_{i,N_2} < q_\alpha^X\}} < s_\epsilon\right) - \mathbb{P}\left(\sum_{i=1}^{N_1} \mathbb{1}_{\{\bar{X}_i - \Delta_{i,N_2} \leq q_\alpha^X\}} < r_\epsilon\right).
$$

Again, we use the advantage that the sum over an indicator function of any subset of iid random variables is binomial distributed. Thereby, the upper bound of summation is the number of trials, the subset of the indicator function defines the success probability and the right-hand side of each inequality indicates the number of success. For $p_{N_2}^+ = \mathbb{P}\left(\bar{X}_i + \Delta_{i,N_2} < q_\alpha^X\right)$ and $p_{N_2}^- = \mathbb{P}\left(\bar{X}_i - \Delta_{i,N_2} \leq q_\alpha^X\right)$ this yields

$$
\mathbb{P}\left(\sum_{i=1}^{N_1} \mathbb{1}_{\{\bar{X}_i + \Delta_{i,N_2} < q_\alpha^X\}} < s_\epsilon\right) - \mathbb{P}\left(\sum_{i=1}^{N_1} \mathbb{1}_{\{\bar{X}_i - \Delta_{i,N_2} \leq q_\alpha^X\}} < r_\epsilon\right)
$$
$$
= \mathcal{B}_{N_1, p_{N_2}^+}\left(s_\epsilon - 1\right) - \mathcal{B}_{N_1, p_{N_2}^-}\left(r_\epsilon - 1\right).
$$

Next, we consider $\mathcal{B}_{N_1, p_{N_2}^+}\left(s_\epsilon - 1\right) - \mathcal{B}_{N_1, \alpha + \frac{\delta}{2}}\left(s_\epsilon - 1\right)$ and $\mathcal{B}_{N_1, \alpha - \frac{\delta}{2}}\left(r_\epsilon - 1\right) - \mathcal{B}_{N_1, p_{N_2}^-}\left(r_\epsilon - 1\right)$ in order to obtain an inequality for our confidence interval. Since the cdf of a binomial distribution is continuously differentiable in p (cf. Lemma B.1), the intermediate value theorem for $\xi \in (p_{N_2}^+, \alpha + \frac{\delta}{2})$ yields

$$
\mathcal{B}_{N_1, \alpha + \frac{\delta}{2}}\left(s_\epsilon - 1\right) - \mathcal{B}_{N_1, p_{N_2}^+}\left(s_\epsilon - 1\right) = \frac{\partial}{\partial \xi}\mathcal{B}_{N_1, \xi}\left(s_\epsilon - 1\right) \cdot \left(\alpha + \frac{\delta}{2} - p_{N_2}^+\right)
$$
$$
= -N_1 \cdot \binom{N_1 - 1}{s_\epsilon - 1} \cdot \xi^{s_\epsilon - 1} \cdot (1 - \xi)^{N_1 - s_\epsilon} \cdot \left(\alpha + \frac{\delta}{2} - p_{N_2}^+\right)
$$
$$
= N_1 \cdot \binom{N_1 - 1}{s_\epsilon - 1} \cdot \xi^{s_\epsilon - 1} \cdot (1 - \xi)^{N_1 - s_\epsilon} \cdot \left(p_{N_2}^+ - \left(\alpha + \frac{\delta}{2}\right)\right).
$$

Since $\binom{N_1 - 1}{s_\epsilon - 1} \cdot \xi^{s_\epsilon - 1} \cdot (1 - \xi)^{N_1 - s_\epsilon}$ describes an ordinary binomial probability, it immediately follows

$$
\left|\mathcal{B}_{N_1, \alpha + \frac{\delta}{2}}\left(s_\epsilon - 1\right) - \mathcal{B}_{N_1, p_{N_2}^+}\left(s_\epsilon - 1\right)\right| \leq N_1 \cdot \left(p_{N_2}^+ - \left(\alpha + \frac{\delta}{2}\right)\right).
$$

Now, the (NF) assumption (i.e. $q_{\alpha,-}^X = q_\alpha^X = q_{\alpha,+}^X$ holds) and Proposition 5.3.2 yield for sufficiently large N_2 and fixed N_1

$$
\left|\mathcal{B}_{N_1, p_{N_2}^+}\left(s_\epsilon - 1\right) - \mathcal{B}_{N_1, \alpha + \frac{\delta}{2}}\left(s_\epsilon - 1\right)\right| \leq \mathcal{O}\left(N_2^{-1/2}\right).
$$

For $\mathcal{B}_{N_1, p_{N_1}^-}\left(r_\epsilon - 1\right) - \mathcal{B}_{N_1, \alpha - \frac{\delta}{2}}\left(r_\epsilon - 1\right)$ the same applies, i.e.

$$
\left|\mathcal{B}_{N_1, p_{N_1}^-}\left(r_\epsilon - 1\right) - \mathcal{B}_{N_1, \alpha - \frac{\delta}{2}}\left(r_\epsilon - 1\right)\right| \leq \mathcal{O}\left(N_2^{-1/2}\right).
$$

In summary, for sufficiently large N_2 we obtain thus

$$\mathbb{P}\left(\left(\bar{X} - \Delta_{N_2}\right)_{(r_\epsilon)} \leq q_\alpha^X \leq \left(\bar{X} + \Delta_{N_2}\right)_{(s_\epsilon)}\right)$$

$$\geq \mathcal{B}_{N_1, p_{N_2}^+}\left(s_\epsilon - 1\right) - \mathcal{B}_{N_1, p_{N_2}^-}\left(r_\epsilon - 1\right)$$

$$\geq \mathcal{B}_{N_1, \alpha + \frac{\delta}{2}}\left(s_\epsilon - 1\right) - \mathcal{B}_{N_1, \alpha - \frac{\delta}{2}}\left(r_\epsilon - 1\right) \pm \mathcal{O}\left(N_2^{-1/2}\right).$$

In order to find a $(1 - \epsilon)$-confidence interval we now seek an s_ϵ and r_ϵ such that

$$\mathcal{B}_{N_1, \alpha + \frac{\delta}{2}}\left(s_\epsilon - 1\right) - \mathcal{B}_{N_1, \alpha - \frac{\delta}{2}}\left(r_\epsilon - 1\right) > 1 - \epsilon$$

holds. In total, we obtain thus

$$\mathbb{P}\left(\left(\bar{X} - \Delta_{N_2}\right)_{(r_\epsilon)} \leq q_\alpha^X \leq \left(\bar{X} + \Delta_{N_2}\right)_{(s_\epsilon)}\right) > 1 - \epsilon \pm \mathcal{O}\left(N_2^{-1/2}\right).$$

$\square$

Remark 5.3.4.
Let us emphasize here some important remarks:

- *Theorem 5.3.3 establishes model free confidence intervals. Thus, σ^2 is in this setup generally unknown. According to that, we cannot use the classical Berry-Esseen Theorem (cf. Theorem 2.4.2) in Proposition 5.3.2 and must apply an advanced version for the sample variance $\hat{\sigma}_i^2$, see e.g. Bentkus and Götze (1996), Theorem 1.1.*
- *At the end of the proof of Theorem 5.3.3 we seek for $\epsilon > 0$ an s_ϵ and r_ϵ such that*

$$\mathcal{B}_{N_1, \alpha + \frac{\delta}{2}}\left(s_\epsilon - 1\right) - \mathcal{B}_{N_1, \alpha - \frac{\delta}{2}}\left(r_\epsilon - 1\right) > 1 - \epsilon$$

holds in order to obtain an $(1 - \epsilon)$-confidence interval. Here, several splitting criteria are possible like for example the statistical standard of $\epsilon/2$ and $1 - \epsilon/2$. In an insurance framework and thus in the SCR case it seems to be advantageous to seek a sharp lower bound. Hence, a splitting such that the lower bound is as close as possible to the Value-at-Risk is sought. Thus, since $VaR(X) = -q_\alpha^X$ holds, we obtain $\mathbb{P}\left(-\widehat{UB}_{N_2, \epsilon} \leq -q_\alpha^X \leq -\widehat{LB}_{N_2, \epsilon}\right) > 1 - \epsilon$ and seek an s_ϵ and r_ϵ based on an $1 - \epsilon/10$ and $(9\,\epsilon)/10$ split.

- *Simulations particularly on these splitting criteria and comparisons to the already existing approach of Lan et al. (2007b) can be found in Section 6.1. Moreover, in Section 6.2 we apply this approach on the insurance model of Hieber et al. (2019).*
- *Further, note that per splitting criteria several s_ϵ and r_ϵ indices can exist. Hence, there exist s_ϵ and r_ϵ couples per splitting criteria which should be preferred since they minimize the confidence interval width compared to other couples. For example in the classical $\epsilon/2$ resp. $1 - \epsilon/2$ splitting we seek especially*

$$\widehat{LB}_{N_2,\epsilon} = \arg\min_{r_\epsilon} \left\{ \mathcal{B}_{N_1, \alpha - \frac{\delta}{2}} (r_\epsilon - 1) \leq \frac{\epsilon}{2} \right\}$$

$$\widehat{UB}_{N_2,\epsilon} = \arg\min_{s_\epsilon} \left\{ \mathcal{B}_{N_1, \alpha + \frac{\delta}{2}} (s_\epsilon - 1) \geq 1 - \frac{\epsilon}{2} \right\}.$$

In this work, we do not optimize over these possible indices and leave this point open for later considerations.

Finally, the following Table 5.1 summarizes the underlying effects which occur in the noise considering confidence interval method.

Table 5.1 Confidence interval indices based on our new proposed approach and the standard method, with and without noise consideration, and different outer and inner settings for fixed $\epsilon = 1\%$, $\delta = 0.05\%$ and $\alpha = 99.5\%$

	without noise		with noise		with noise		CLT Impact	
N_1	$r_{1\%}$	$s_{1\%}$	$r_{1\%}$	$s_{1\%}$	$r_{1\%}$	$s_{1\%}$	$\frac{1}{N_1} \sum_{i=1}^{N_1} \Delta_{i,N_2}$	
10^3	988	1000	984	1000	984	1000	0.0262	0.2465
10^4	9926	9968	9901	9987	9901	9987	0.0261	0.2433
10^5	99442	99558	99178	99790	99178	99790	0.0264	0.2467
N_2	/	/	10^3		10		10^3	10

First, compared to the standard confidence intervals without noise consideration the noise considering intervals lead—not surprisingly—to larger indices according to the consideration of the inner error. Hence, they lead to wider intervals which in turn capture the quantile also in a noisy environment with few inners, cf. Figure 5.1 and the corresponding discussion. Second, we immediately see that the underlying noise level resp. the number of inner scenarios has no influence on the indices $r_{1\%}$ and $s_{1\%}$. This effect holds in general since only the number of outer scenarios is important because they influence the size of the underlying binomial distribution directly. The CLT impact, however, addresses the underlying noise level and thus the choice of N_2. Here, as third we have that a larger amount of inners, i.e. a less noisy simulation, leads to a smaller CLT surplus and thus in the end to tighter intervals. However, for a quantile level of $\alpha = 99.5\%$ a simulation of at least $N_1 \geq 10^4$ outer scenarios or a smaller CLT level δ is needed, such that the upper bound of the confidence interval, i.e. the index $s_{1\%}$, can be distinguished from the end of the underlying binomial distribution. This problem occurs since $\alpha = 99.5\%$ implies a rather extreme tail quantile level. Thus, a minimum number of outer scenarios is needed for a clear distinction.

Numerical Analysis

6

To illustrate results, this chapter introduces an academic as well as a very simplified insurance specific setup. The academic setup allows a derivation of an analytical solution for $\mathbb{E}[G(X)]$ in the moment- and q_α^X in the quantile-based case. Hence, such an example is especially relevant for our almost sure considerations because the illustration of a.s. rates requires at least an analytical solution. In the insurance specific part we introduce, first, an ALM model (cf. Hieber et al. (2019)) and describe its components. Second, we use this model to compare our confidence interval method to the already existing one of Lan et al. (2007b) in a practical setup. All simulation studies were performed on a standard laptop (processor: Intel core i7-1165G7, with 2.8 GHz, RAM: 16 GB) and all implementations were carried out in MATLAB 2018b.

6.1 Academic Test Examples

In the following, we distinguish between the problem statements (i.e. moment-based, quantile-based and confidence intervals) and verify the theoretical rates of convergence and confidence interval widths based on various academic test examples.

Almost Sure Convergence Moment-Based

To validate the a.s. rates in the moment-based case we consider two different problem formulations, first, the commonly known large loss problem, (also investigated in Hong and Juneja (2009), Gordy and Juneja (2010), Broadie et al. (2015), Liu et al. (2022)), and, second, a polynomial case with $G(X) = X^2$. Based on sufficient

© The Author(s), under exclusive license to Springer Fachmedien Wiesbaden GmbH, part of Springer Nature 2024
M. Klein, *Nested Simulations: Theory and Application*, Mathematische Optimierung und Wirtschaftsmathematik I Mathematical Optimization and Economathematics, https://doi.org/10.1007/978-3-658-43853-1_6

conditions on V as well as X both cases cover different vanishing bias rates, i.e. $\beta_2 = 1/2$ and the faster one $\beta_2 = 1$. Consequently, they are predestined to numerically validate the respective rates from Table 3.1. For the large loss problem the subsequent settings will be considered:

Example 6.1.1.
Let $Z \sim \mathcal{U}_{(0,1)}$, $Y \sim \mathcal{N}(0,1)$ and $V = Z + Y$, whereby Y is independent of Z. Then, we obtain

$$X = \mathbb{E}[V|Z] = Z,$$

i.e. $X \sim \mathcal{U}_{(0,1)}$, according to the independence and measurability properties of Y and Z. Here, we consider $G(X) = \mathbb{1}_{\{X > 0.8\}}$, which immediately yields

$$\mathbb{E}[G(X)] = \mathbb{E}\left[\mathbb{1}_{\{X > 0.8\}}\right] = \mathbb{P}(X > 0.8) = 0.2.$$

To illustrate the almost sure behaviour of $\left\{\bar{\gamma}_{N_1(N),N_2(N)}\right\}$ Figure 6.1 depicts the approximation error, i.e. $|\bar{\gamma}_{N_1(N),N_2(N)} - \gamma|$, for 200 sample paths on a $\log_2$-scale $N = 2^5, \ldots, 2^{40}$ and compares them with various theoretical rates, cf. corresponding solid dashed line. Thereby, two different budget sequences and vanishing bias rates β_2 were investigated in more detail.

Since $\|\mathbb{E}\left[|V|^p \mid Z\right]\|_{L^\infty} < \infty$ holds in this example, it is known from prior analysis that $(A1,\infty)$ (cf. Proposition 3.2.1) and $(VB,\frac{1}{2})$ (cf. Proposition 3.2.7) applies. Hence, all assumptions in Theorem 3.4.4 are satisfied and according to Table 3.1 we obtain for $p = 6$ and $\beta_2 = 1/2$ an optimal rate of (almost) $N^{-1/4}$, with $\beta_1 = 1/2$ and $r = 1/2$, i.e. the underlying budget sequence is $N_1(N) = N_2(N) = N^{1/2}$. Consequently, in accordance with theory it can be observed that nearly all sample paths (out of 200) satisfy the optimal rate $N^{-1/4}$.

Next, we analyze the $N^{-1/3}$ RMSE rate of Hong and Juneja (2009) and Gordy and Juneja (2010) closer. Hence, we apply, first, their budget sequence, i.e. we choose $r = 1/3$ which yields to $N_1(N) = N^{2/3}$ and $N_2(N) = N^{1/3}$ and second, their fast bias rate $\beta_2 = 1$, i.e. (VB,1). Then, according to Table 3.1 we also obtain an almost sure optimal rate of $N^{-1/3}$. Figure 6.1 (bottom) indicates that immediately because nearly all sample paths satisfy the proposed rate of convergence.

Second, in the polynomial case the following simulation framework will be considered:

Example 6.1.2.
We investigate $G(X) = X^2$ and choose the outer scenarios Z as standard normal distributed, i.e. $Z \sim \mathcal{N}(0,1)$, and the inner scenarios as lognormal distributed with parameters $\mu = 0$ and $\sigma^2 = Z^2/t$, i.e. $V|Z \sim \mathcal{LN}(0, Z^2/t)$, for $t > 2p$.

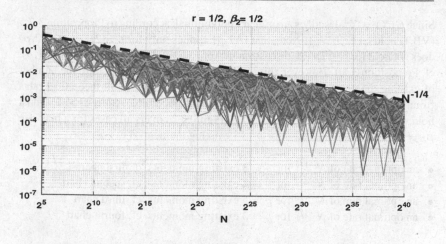

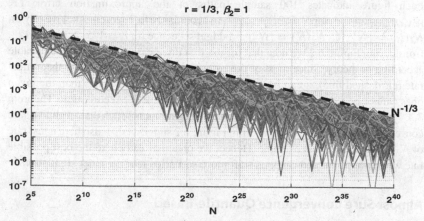

Figure 6.1 Plots of 200 sample paths of $|\bar{\gamma}_{N_1(N),N_2(N)} - \gamma|$ and their corresponding complexities (dashed line) for $r = 1/2$, $\beta_2 = 1/2$ (top) and $r = 1/3$, $\beta_2 = 1$ (bottom) and different sample sizes $N = 2^5, \ldots, 2^{40}$

Thereby, p denotes the number of existing moments in (A1, p). For $t > 2$, we obtain an analytical solution for γ (cf. Appendix A.1), i.e.

$$\mathbb{E}\left[G(X)\right] = \mathbb{E}\left[G\left(\mathbb{E}[V|Z]\right)\right] = \frac{1}{\sqrt{1 - 2 \cdot \frac{1}{t}}}.$$

Since $G(X) = X^2$ describes an ordinary polynomial, according to Proposition 3.2.5 (VB,1) holds, i.e. $\beta_2 = 1$. Note, that (A1,p) applies only for $t > 2p$ since (A2,p) does not hold for $t \leq 2p$ (cf. discussion in Appendix A.1) and thus also not (A1,p), cf. Proposition 3.1.1. However, this implies that it is possible to control the number of existing moments in this example. Hence, a verification of all theoretical rates from Table 3.1 for different p and fixed $\beta_2 = 1$ is possible. Since $t > 2p$ must hold, we set $t = 2 \cdot p + \epsilon$ with $\epsilon = 10^{-3}$. Figure 6.2 covers all cases for a bias rate $\beta_2 = 1$ and an additional case for $p = 3$ existing moments, i.e.

- an optimal rate of $N^{-1/5}$ for $p = 3$ existing moments, cf. first chart
- an optimal rate of $N^{-2/7}$ for $p = 4$ existing moments, cf. second chart
- an optimal rate of $N^{-1/3}$ for $p = 5$ existing moments, cf. third chart
- an optimal rate of $N^{-1/3}$ for $p = 3$ existing moments, cf. fourth chart

Each figure indicates 100 sample paths of the approximation error, i.e. $|\bar{\gamma}_{N_1(N), N_2(N)} - \gamma|$, for $N = 2^3, \ldots, 2^{29}$. In accordance to Theorem 3.4.4 we rely on $N_1(N) = \lfloor N^{1-r} \rfloor$, $N_2(N) = \lfloor N^r \rfloor$ as budget sequence for $r = 1/5, 2/7, 1/3, 1/3$. Considering Figure 6.2 we note that for the first three cases all rates are achievable according to theory. Since most of the paths are extremely close to the theoretical rate (cf. dashed line), it can be, indeed, concluded that better rates than the given ones cannot be expected for the chosen budget sequences. According to Theorem 3.4.4 we cannot obtain $r = 1/3$ for $p = 3$ existing moments such that an almost sure convergence holds. But the last chart in Figure 6.2 suggests that also here a fast rate of $N^{-1/3}$ applies in this example. Hence, we obtain already here the best possible rate we can hope for and can thus shift more scenarios into the inner simulation.

Almost Sure Convergence Quantile-Based

As in the previous part we start with a rather simple example and simulation framework in accordance to a previous theoretical consideration:

Example 6.1.3.
Let $Z \sim \mathcal{N}(0, 1)$, $Y \sim \mathcal{N}(0, 1)$ and $V = Z + Y$, whereby Y is independent of Z. Then, we immediately obtain

$$X = \mathbb{E}[V|Z] = Z,$$

i.e. $X \sim \mathcal{N}(0, 1)$, according to the independence of Y and Z. Moreover, based on the definition of ϵ_{N_2}, we get

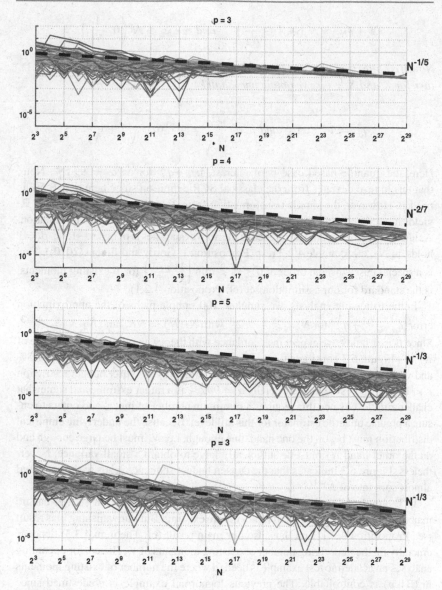

Figure 6.2 Plots of 100 sample paths of $|\bar{\gamma}_{N_1(N),N_2(N)} - \gamma|$ and their corresponding rates (dashed line) for $p = 3$ (first), $p = 4$ (second), $p = 5$ (third), $p = 3$ (fourth) existing moments and different sample sizes $N = 2^3, \ldots, 2^{29}$

$$X + \epsilon_{N_2} \sim \mathcal{N}\left(0, 1 + \frac{1}{N_2}\right) \quad and \quad \epsilon_{N_2} \sim \mathcal{N}\left(0, \frac{1}{N_2}\right).$$

Now, it is quite straightforward to determine the underlying quantiles. Let $\alpha \in (0, 1)$, then for X and $X + \epsilon_{N_2}$ we obtain immediately

$$q_\alpha^X = \Phi^{-1}(\alpha) \quad and \quad q_\alpha^{X+\epsilon_{N_2}} = \Phi^{-1}(\alpha) \cdot \sqrt{1 + \frac{1}{N_2}}.$$

Here, the quantile-based case is of interest, i.e. $-q_\alpha^X$ for an $\alpha = 99.5\%$. Note, that this in turn deviates from the classical SCR definition since here $\alpha = 0.5\%$ is needed. However, the simple transformation from X to the loss $c - X$, for $c \in \mathbb{R}$, yields immediately in a interpretation of the introduced SCR problem formulation. As already mentioned, the difference of the above quantiles is of order -1, i.e. $\beta = 1$ holds, in N_2, cf. Example 4.2.3 resp. Proposition 1, Gordy and Juneja (2010)). This is a special case since an order of $0 \le \beta < \min\left\{\frac{1}{2}, \frac{p-1}{p+1}\right\}$ for p existing moments, is the standard for our assumption set (cf. Proposition 4.2.1).

In the following analysis, we simulate 100 sample paths of the approximation error $\left|\hat{q}_{\alpha, N_1}^{X+\epsilon_{N_2}} - q_\alpha^X\right|$ for $N = 2^5, \ldots, 2^{40}$ with $N_1(N) = N^{2/3}$ and $N_2(N) = N^{1/3}$. Since $\beta = 1$ holds, we expect in accordance with theory an (almost) optimal rate of $N^{-1/3}$. Figure 6.3 considers three different quantile levels $\alpha = 70\%, 90\%, 99.5\%$ and verifies the rates only approximately according to the strict budget restriction $r < 1/(1 + 2\beta)$. Note, that especially $\alpha = 99.5\%$ is a rather extreme and somewhat challenging level since this requires a minimum number of inners and outers to ensure a precise quantile estimator for this tail level. Because the underlying empirical distribution must be, on the one hand, fine enough, i.e. N_1 must be large enough and on the other hand N_2 must be sufficiently large to obtain a small variance. Nevertheless, Figure 6.3 indicates that the chosen budget sequence leads to the proposed almost sure rate of $N^{-1/3}$.

In Example 6.1.3, $V \sim \mathcal{N}(0, 2)$ applies and thus assumption (B1,p) has no major influence because here all moments are existing. But our vanishing bias result (see Proposition 4.2.1) and thus also our main result (cf. Theorem 4.3.5) depends crucially on the underlying number of existing moments of V. Hence, for upcoming analyses an academic test example is needed where the number of existing moments in (B1,p) is controllable. The previous lognormal example is predestined since according to Appendix A.2 a condition on t can be derived such that $||V||_{L^p} < \infty$ holds. Example 6.1.4 repeats the simulation framework and derives the needed analytical quantile q_α^X.

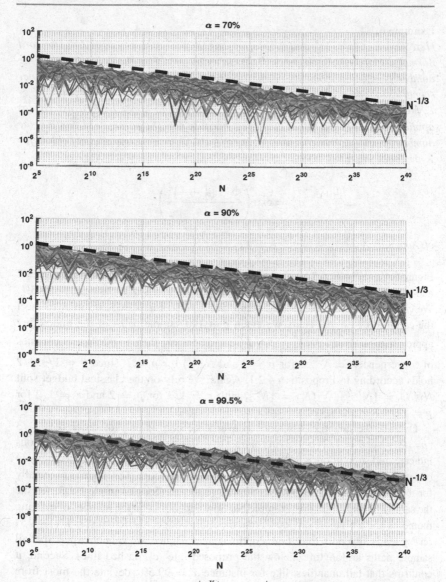

Figure 6.3 Plots of 100 sample paths of $\left| \widehat{q}_{\alpha, N_1}^{X + \epsilon_{N_2}} - q_\alpha^X \right|$ and their corresponding rate (dashed lined) for $\alpha = 70\%$ (top), $\alpha = 90\%$ (middle), $\alpha = 99.5\%$ (bottom) and different sample sizes $N = 2^5, \ldots, 2^{40}$

Example 6.1.4.
Here we investigate $-q_\alpha^X$ *and choose, once again, the outer scenarios* Z *as standard normal distributed, i.e.* $Z \sim \mathcal{N}(0, 1)$. *In order to reflect the dependency of the number of existing moments on the a.s. rates of convergence we take the inner scenarios lognormal distributed with parameters* $\mu = 0$ *and* $\sigma^2 = Z^2/t$, *i.e.* $V|Z \sim \mathcal{LN}(0, Z^2/t)$, *for* $t > p^2$. *If* $t > p^2$ *holds, according to Appendix A.2, also* $(B1, p)$ *applies for* p. *This is a rather nice effect because now the parameter* t *controls the number of existing moments from* V. *Let* $\alpha \in (0, 1)$, *then we immediately obtain for* X *the following quantile*

$$q_\alpha^X = \exp\left(\frac{\Phi^{-1}\left(\frac{1}{2} + \frac{\alpha}{2}\right)^2}{2t}\right),$$

cf. Appendix A.2.

Figure 6.4 and Figure 6.5 cover now all analyses for the two important cases of $p = 2$ and $p = 3$ existing moments for different quantile levels $\alpha = 70\%, 90\%, 99.5\%$. We choose $t = 5$ for the $p = 2$ setup and $t = 10$ for $p = 3$, in order to meet the condition $t > p^2$. Here, we again plot 100 sample paths of the underlying approximation error, i.e. $\left|\tilde{q}_{\alpha, N_1}^{X+\epsilon_{N_2}} - q_\alpha^X\right|$, and compare them to the theoretical rates of convergence, i.e. $N^{-1/5}$ for $p = 2$ and $N^{-1/4}$ for $p = 3$, since $\beta = 1/3, 1/2$ holds according to Proposition 4.2.1. Again, we rely on the classical budget split $N_1(N) = \lfloor N^{1-r} \rfloor$, $N_2(N) = \lfloor N^r \rfloor$, with $r = 1/3$ for $p = 2$ and $r = 1/2$ for $p = 3$ (cf. Remark 4.3.6).

Overall, the quantile difference convergence here is only of order $\beta = 1/3$ ($p = 2$) resp. $\beta = 1/2$ ($p = 3$) in contrast to $\beta = 1$ before. Indeed, both simulations indicate and support our results that more existing moments p lead immediately to better rates and thus, also to nicer convergence behaviours. This applies especially for the extreme case $\alpha = 99.5\%$. Similar to the previous Example 6.1.3, we see the same effects here in this example but now depending on the number of existing moments and a bit more obvious. First, the sharp budget restriction $r < 1/(1+2\beta)$ on N_2 leads also only to (almost) optimal rates. Hence, we cannot expect that all sample paths are anytime below the proposed rate (cf. dashed line). Second, it confirms that tail quantiles, like for instance $\alpha = 99.5\%$, deviate the most from the theoretical rates. Hence, the corresponding constant seems to play a crucial role here. According to Theorem 4.3.5 the constant depends mainly on the density of X, i.e. $F_X'(q_\alpha^X)$. But in this simulation framework we have not considered any constant effects and focused mainly on the pure rate of convergence without a constant factor.

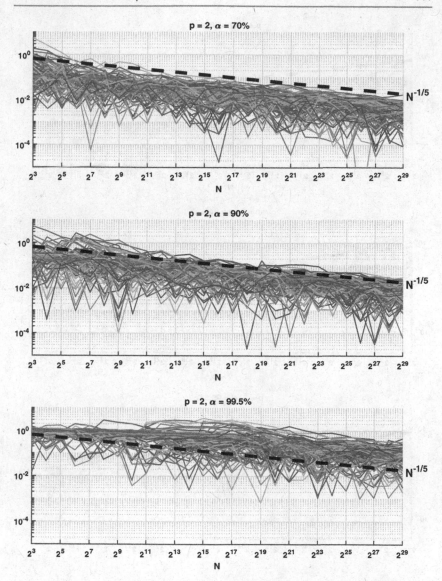

Figure 6.4 Based on $p = 2$ existing moments plots of 100 sample paths of $\left|\widehat{q}_{\alpha,N_1}^{X+\epsilon_{N_2}} - q_\alpha^X\right|$ and their corresponding rate (dashed lined) for $\alpha = 70\%$ (top), $\alpha = 90\%$ (middle), $\alpha = 99.5\%$ (bottom) and different sample sizes $N = 2^3, \ldots, 2^{29}$.

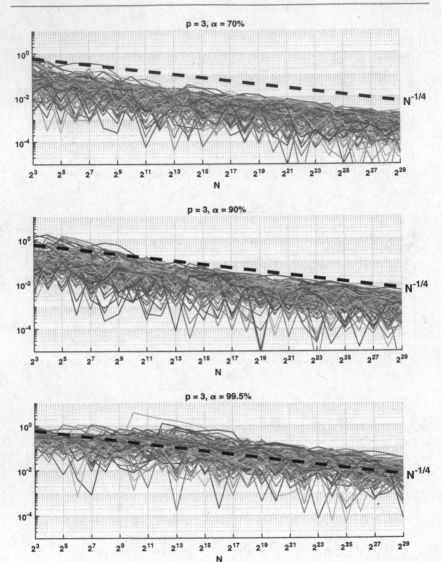

Figure 6.5 Based on $p = 3$ existing moments plots of 100 sample paths of $\left| \widehat{q}_{\alpha, N_1}^{X + \epsilon_{N_2}} - q_{\alpha}^{X} \right|$ and their corresponding rate (dashed lined) for $\alpha = 70\%$ (top), $\alpha = 90\%$ (middle), $\alpha = 99.5\%$ (bottom) and different sample sizes $N = 2^3, \ldots, 2^{29}$

Confidence Intervals

Here, non parametric confidence intervals based on the presented method in Chapter 5 will be derived and compared to the so far only existing method of Lan et al. (2007b).

We rely on the previous simulation framework from Example 6.1.2 and obtain $-q_\alpha^X \approx -1.482859$ for $t = 10$ and $\alpha = 99.5\%$ (see Appendix A.2). The overall coverage probability of the confidence interval is fixed to $\epsilon = 1\%$, as CLT coverage probability we choose $\delta = \delta_L = 1 - (1 - \epsilon_{in})^{1/N_1}$ for comparison reasons. In Lan et al. (2007b) the overall coverage probability ϵ must be divided into two separate coverage probabilities, namely one for the outer simulation and one for the inner. In this comparing simulation we divide it equally and take thus $\epsilon_{out} = \epsilon_{in} = 0.5\%$. However, it should be highlighted that our approach seems to be a bit more intuitive, especially the parametrization part because only the logical parameters quantile level α, CLT probability δ and coverage probability ϵ have to be initialized. Furthermore, it is evident that in our interval method δ seems to be the crucial parameter according to its influence on the noise consideration. Contrary, in Lan et al. (2007b) ϵ_{in} is the crucial parameter with its severe effect on ϵ_{out} and δ_L, but indeed it is unclear which choice is appropriate. All upcoming simulations rely on $N_1 = 10^4$ fixed outer and $N_2 = 2^3, \ldots, 2^{15}$ inner scenarios.

In Figure 6.6 each solid line indicates the median widths of the three existing non parametric confidence interval methods over 1000 simulation runs. The additional vertical intervals around each line describe the difference between the occurred maximal resp. minimal width and the corresponding median of each simulation set. Thereby, the classical symmetrical two-sided confidence interval split of ϵ was used, i.e. $\epsilon/2$ and $1 - \epsilon/2$. We see immediately that the confidence intervals without noise consideration lead to the tightest intervals. But, as already seen, they misclassify $-q_\alpha^X$ for few inner scenarios, cf. Figure 5.1. Both noise considering methods lead to no misclassifications, cf. Table 6.1. It immediately turns out that for small N_2 (i.e. $N_2 \le 2^5$) our new approach yields significantly tighter intervals in comparison to Lan et al. (2007b), first by smaller median values and second by a significantly smaller difference between the maximal and minimal width. For large N_2 and thus a low noise level both approaches yield similar interval widths.

The previous simulation concentrates on the classical two-sided confidence interval case with a symmetrical coverage probability splitting of $\epsilon/2$ and $1 - \epsilon/2$. According to the risk capital interpretation of the SCR for insurance companies it is

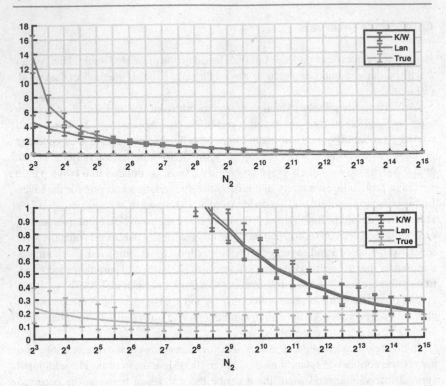

Figure 6.6 Numerical simulation study of an analytical nested Monte Carlo model with $N_1 = 10^4$ fixed and rising $N_2 = 2^3, \ldots, 2^{15}$. Median widths (solid line) of the occurring symmetric confidence intervals over 1000 simulations for $\alpha = 99.5\%$ and $\epsilon = 1\%$. The additional vertical error bars indicate the corresponding difference between the maximum resp. minimum width and the median

even more important to obtain sharp lower bounds for the quantile instead of symmetric bounds. Therefore, in the following analysis we investigate an asymmetric case, i.e. we use the following splitting $(9\,\epsilon)/10$ for the upper bound and $1 - \epsilon/10$ for the lower bound. The rest of the simulation setup is identical to the previous one. Figure 6.7 indicates a nearly analogous picture as before. It turns out that the median widths of our new proposed approach lead to tighter intervals in noisy environments (i.e. $N_2 \leq 2^{10}$). For $N_2 \geq 2^{11}$, however, both noise considering methods perform nearly equally and the differences are only marginal.

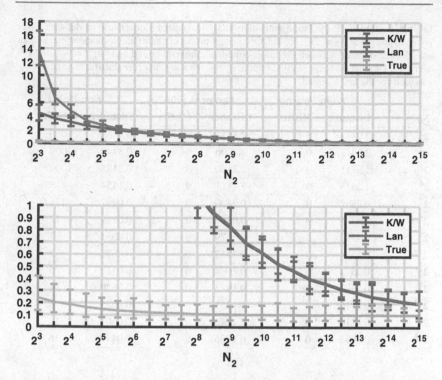

Figure 6.7 Numerical simulation study of an analytical nested Monte Carlo model with $N_1 = 10^4$ fixed and rising $N_2 = 2^3, \ldots, 2^{15}$. Median widths (solid line) of the occurring asymmetric confidence interval widths over 1000 simulations for $\alpha = 99.5\%$ and $\epsilon = 1\%$. Thereby, we used for the lower bound an order statistic level of $1 - \frac{\epsilon}{10}$ and for the upper bound $\frac{9\epsilon}{10}$. The additional vertical error bars indicate the corresponding difference between the maximum resp. minimum width and the median

How well or poorly the respective confidence interval methods precisely perform for a sharp left boundary, is depicted in Table 6.1. Here, we summarize the median widths between the respective left boundary and the analytical quantile $-q_\alpha^X$ for $\alpha = 99.5\%$ again over 1000 simulation runs per $N_2 = 2^3, \ldots, 2^{15}$ and fixed $N_1 = 10^4$. It is immediately clear that for the noise considering intervals our approach outperforms the method of Lan et al. (2007b) in all noise scenarios and leads thus to sharper left-hand bounds. Especially, for $N_2 < 2^7$ significant differences are noticeable. Again, the non-noisy intervals (cf. David and Nagaraja (2004)) lead to the sharpest lower bound but just as in the symmetrical case to a huge amount of

Table 6.1 Numerical simulation study of an analytical nested Monte Carlo model with $N_1 = 10^4$ fixed and rising $N_2 = 2^3, \ldots, 2^{15}$. Occurring number of misfits and corresponding median levels of the distance between the lower confidence interval bound and $-q_\alpha^X \approx -1.482859$ over 1000 simulations for $\alpha = 99.5\%$ and $\epsilon = 1\%$. Thereby, an order statistic level of $1 - \frac{\epsilon}{10}$ was used for the lower bound

N_2	Klein / Werner		Lan et al. (2007b)		Non-Noisy	
	Distance	**Misfits**	**Distance**	**Misfits**	**Distance**	**Misfits**
2^3	3.9835	0	12.8430	0	0.5453	1000
2^4	2.6913	0	4.3206	0	0.3449	1000
2^5	1.8211	0	2.2650	0	0.2185	1000
2^6	1.2675	0	1.4111	0	0.1454	815
2^7	0.8867	0	0.9400	0	0.0997	171
2^8	0.6334	0	0.6569	0	0.0769	22
2^9	0.4562	0	0.4679	0	0.0631	3
2^{10}	0.3354	0	0.3432	0	0.0573	4
2^{11}	0.2543	0	0.2604	0	0.0557	6
2^{12}	0.1917	0	0.1973	0	0.0520	14
2^{13}	0.1515	0	0.1551	0	0.0525	8
2^{14}	0.1210	0	0.1245	0	0.0507	10
2^{15}	0.0995	0	0.1035	0	0.0499	16

misfits, i.e. the underlying interval is too narrow especially for $N_2 \leq 2^9$ and thus, does not cover the quantile at hand (cf. motivation in Chapter 5).

6.2   Insurance Model

In the following, we rely on the insurance model proposed by Hieber et al. (2019). Here, the authors investigate participating life insurance portfolios in a homogeneous and especially a heterogeneous setup. In our context we are interested in a nested Monte Carlo framework and the behaviour of confidence intervals for the Value-at-Risk and not on insurance peculiarities. Thus, we restrict the following analysis to the homogeneous case.

Example 6.2.1.
For the SCR derivation two time horizons are necessary. First, the risk horizon $t = t_0, \ldots, t_1$ with its underlying physical probability measure $\mathbb{P}$ and second, the

valuation horizon $t = t_1, \ldots, t_K$ *with assumed risk neutral measure* $\mathbb{Q}$. *Hieber et al.* (2019) *consider a risk-free security* $\{r_t\}_{t \geq 0}$ *and a risky asset* $\{S_t\}_{t \geq 0}$. *The risky asset is given by the following standard stochastic differential equation*

$$dS_t = \mu \, S_t \, dt + \sigma_1 \, S_t \, dW_t^{\mathbb{P}},$$

whereby σ_1 *is a positive constant,* $\mu \in \mathbb{R}$ *and* $W_t^{\mathbb{P}}$ *denotes a standard Brownian motion under* $\mathbb{P}$. *For the risk-free asset* $\{r_t\}_{t \geq 0}$ *they take a Vasicek interest rate model, i.e.*

$$dr_t = \kappa \, (\theta - r_t) \, dt + \sigma_2 \, d\widetilde{W}_t^{\mathbb{P}},$$

with mean reversion speed parameter $\kappa > 0$, *long term mean* $\theta \in \mathbb{R}$, $\sigma_2 > 0$ *and a standard Brownian motion* $\widetilde{W}_t^{\mathbb{P}}$. *Here, correlated Brownian motions are assumed and thus* $dW_t^{\mathbb{P}} \cdot d\widetilde{W}_t^{\mathbb{P}} = \rho \, dt$ *holds for* $\rho \in [-1, 1]$. *The numéraire process* B_t *is straightforward and given by*

$$B_t = \exp\left(\int_0^t r_s \, ds \right).$$

Since one major task is the reevaluation of the assets and liabilities, we need a change of measure to an equivalent risk neutral measure $\mathbb{Q}$. *Hence, the risky asset dynamic under* $\mathbb{Q}$ *is given by*

$$dS_t = r_t \, S_t \, dt + \sigma_1 \, S_t \, dW_t^{\mathbb{Q}},$$

where $W_t^{\mathbb{Q}}$ *denotes a standard Brownian motion under* $\mathbb{Q}$. *The risk-free security under* $\mathbb{Q}$ *is then*

$$dr_t = \kappa \, (\theta^* - r_t) \, dt + \sigma_2 \, d\widetilde{W}_t^{\mathbb{Q}},$$

with $\theta^* := \theta - \frac{\lambda \cdot \sigma_2}{\kappa}$ *for an assumed constant market price of the interest rate risk* λ. *These are the dynamics for the underlying risk factors. In order to approximate the SCR resp. the loss distribution an insurance portfolio and thus assets and liabilities are needed to determine the Basic own Funds resp.* X, *cf. (1.9). Hieber et al. (2019) suppose that the insurer invests a constant share* η *in the risk-free security and thus* $1 - \eta$ *into the risky asset. The investment portfolio is then defined as*

$$A_{t_k} = A_{t_0} \cdot \exp\left(\int_{t_0}^{t_k} r_s \, ds - \frac{1}{2}(1 - \eta)^2 \sigma_1^2 \, t_k + (1 - \eta)\sigma_1 \, W_{t_k}^{\mathbb{Q}} \right)$$

for $k = 1, \ldots, K$. As already mentioned, Hieber et al. (2019) are investigating participating life insurance contracts. Therefore, the considered policyholder account is as follows

$$L_{t_k} = L_{t_{k-1}} \cdot \max \left\{ \exp(g), \left(\frac{A_{t_k}}{A_{t_{k-1}}} \right)^{\gamma_{t_k}} \right\},$$

for $k = 1, \ldots, K$, with a guaranteed minimum return g and a participation rate $\gamma_{t_k} = \gamma \cdot \mathbb{1}_{\{A_{t_k} \geq A_{t_{k-1}}\}} + \mathbb{1}_{\{A_{t_k} < A_{t_{k-1}}\}}$ on the risky asset with $\gamma \in [0, 1]$.

In this nested Monte Carlo framework the simulation then takes place as follows:

1. Simulate N_1 underlying risk scenarios $\left(r_t^{(i)}, \int_{t_{k-1}}^{t_k} r_s^{(i)} ds, \right.$

   $\left. \ln(\tilde{S}_{t_k}^{(i)} / \tilde{S}_{t_{k-1}}^{(i)}) \right)_{\substack{i=1,\ldots,N_1 \\ k=0,1}}$ under $\mathbb{P}$, based on Proposition 10 of Hieber et al. (2019).
   Note, that $\tilde{S}_t := S_t / B_t$ denotes the discounted asset value.

2. Simulate the $N_1 \cdot N_2$ risk scenarios $\left(r_{t_k}^{(i,j)}, \int_{t_{k-1}}^{t_k} r_s^{(i,j)} ds, \right.$

   $\left. \ln(\tilde{S}_{t_k}^{(i,j)} / \tilde{S}_{t_{k-1}}^{(i,j)}) \right)_{\substack{i=1,\ldots,N_1 \\ j=1,\ldots,N_2 \\ k=1,\ldots,K}}$ under $\mathbb{Q}$.

3. Compute the assets $\left\{ A_{t_k}^{(i,j)} \right\}_{\substack{i=1,\ldots,N_1 \\ j=1,\ldots,N_2 \\ k=1,\ldots,K}}$, the policyholder accounts

   $\left\{ L_{t_K}^{(i,j)} \right\}_{\substack{i=1,\ldots,N_1 \\ j=1,\ldots,N_2 \\ k=1,\ldots,K}}$ under $\mathbb{Q}$ and the numéraire $\left\{ B_{t_k}^{(i,j)} \right\}_{\substack{i=1,\ldots,N_1 \\ j=1,\ldots,N_2 \\ k=1,\ldots,K}}$.

4. Approximate the conditional expectation value (1.9) at t_1 for each outer scenario, i.e. for $i = 1, \ldots, N_1$:

$$\bar{X}_i = \frac{1}{N_1} \sum_{j=1}^{N_2} \left(A_{t_K}^{(i,j)} - L_{t_K}^{(i,j)} \right) \cdot \frac{B_{t_K}^{(i,j)}}{B_{t_1}^{(i,j)}}.$$

Totally in line with Hieber et al. (2019), we make the following assumptions on the underlying financial market, cf. Table 6.2.

Furthermore, as contract parameters we choose a yearly time grid with maturity date $T = 50$, a risk free investment rate of $\eta = 0.9$, a participation rate $\gamma = 0.9$, a guaranteed minimum return $g = 0.458\%$, an initial investment portfolio of $A_0 = 100$ Mio. € and an initial premium of $P_0 = 90$ Mio. €.

Table 6.2 Parameter set for the underlying financial market under the real world measure $\mathbb{P}$ and the risk neutral measure $\mathbb{Q}$

	risky asset	Vasicek interest rate model
Under $\mathbb{P}$	$\mu = 7\%, \sigma_1 = 20\%$	$r_0 = 1.15\%, \kappa = 30\%, \theta = 1.9\%, \sigma_2 = 20\%, \lambda = -23\%, \rho = 15\%$
Under $\mathbb{Q}$	$\sigma_1 = 20\%$	$r_0 = 1.15\%, \kappa = 30\%, \theta^* = 3.05\%, \sigma_2 = 20\%, \rho = 15\%$

Confidence Intervals SCR Case

This section is intended to compare our noise considering interval method with the existing one of Lan et al. (2007b) based on the previously introduced simple insurance ALM model. Hence, we do not recap any regulatory resp. insurance specific specialties besides the already mentioned ones in Chapter 1. For further details on this topic we refer to McNeil et al. (2015) as well as the dissertation of Krah (2020). Here, we are interested in confidence intervals for $\widehat{q}_{0.5\%}^{\mathrm{BoF}_{t_1}}$. For $N_1 = 10^4$ outers and $N_2 = 10^4$ inners we obtain thus an estimated target value of $\widehat{q}_{0.5\%}^{\mathrm{BoF}_{t_1}} \approx 2.8689$ Mio. €. The positive $\widehat{q}_{0.5\%}^{\mathrm{BoF}_{t_1}}$ indicates immediately that no equity injections are needed to fulfill the underlying SCR requirements. Since the profit scenarios, i.e. X, rely on a nested simulation, we are thus interested in applying noise considering confidence interval methods around q_{α}^{X} to capture the occurring noise level ϵ_{N_2}, especially for noisy environments. Figure 6.8 indicates the underlying profit distribution according to Example 6.2.1. The solid red line represents the already addressed quantile estimator $\widehat{q}_{0.5\%}^{\mathrm{BoF}_{t_1}} \approx 2.8689$ Mio. €. The dashed blue lines reflect the corresponding non-noise considering confidence intervals for $\widehat{q}_{0.5\%}^{\mathrm{BoF}_{t_1}}$, cf. David and Nagaraja (2004), whereas the black solid resp. dashed magenta lines indicate the noise considering intervals. Thereby, the black lines represent the intervals according to our proposed approach (see Chapter 5) and the dashed magenta lines indicate the intervals from Lan et al. (2007b). Since in the SCR framework especially sharp lower bounds are needed, we consider – again – an asymmetric splitting for all methods. Based on this single observation, Figure 6.8 shows the analogous trend like in the academic test example part before, namely that the non-noise considering intervals lead to the tightest intervals with the negative effect of misclassifications (cf. Table 6.1) and that in the noise considering case our confidence interval width is tighter compared to the one from Lan et al. (2007b).

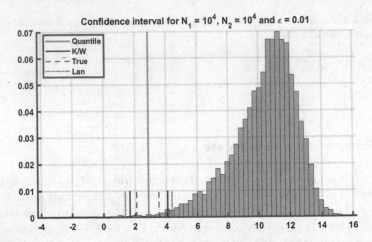

Figure 6.8 BoF_{t_1} histogram based on $N_1 = 10^4$ outer and $N_2 = 10^4$ inner scenarios. The red solid line reflects the estimated quantile, i.e. $\widehat{q}_{0.5\%}^{\text{BoF}_{t_1}} \approx 2.8689$. The blue dashed lines represent the occurring confidence interval without noise considerations, whereas the black and magenta lines represent the noise considering confidence intervals from Klein and Werner (2023a) resp. Lan et al. (2007b) for $\epsilon = 1\%$

Next, we analyze in particular the widths of the noise considering confidence intervals for a fixed number of $N_1 = 10^4$ outer scenarios and a varying number of inner scenarios $N_2 = 10, 10^2, 10^3, 10^4$. Summarized, we seek tight noise considering confidence intervals around the key value $\widehat{q}_{0.5\%}^{\text{BoF}_{t_1}}$. Thereby, we additionally investigate the effect of the underlying CLT level δ resp. δ_L on the interval widths. Lan et al. (2007b) split the confidence level $\epsilon = \epsilon_{out} + \epsilon_{in}$ and consider thus two error sources in their approach uniquely, namely an outer ϵ_{out} and an inner ϵ_{in} source. Since $\delta_L = 1 - (1 - \epsilon_{in})^{1/N_1}$ holds, the inner error source varies in the following investigations and influences the CLT level δ_L directly. Hence, different inner and according to $\epsilon = \epsilon_{out} + \epsilon_{in}$ also outer error source combinations will be tested in the approach of Lan et al. (2007b). For comparison reasons we set the CLT level in our framework to $\delta = \delta_L$. Already here a huge advantage of our approach turns out because the error splitting in Lan et al. (2007b) and consequently its choice lead definitely to a higher parameterization complexity.

In Figure 6.9 we consider a confidence level of $\epsilon = 0.1\%$ and a varying inner error source of $\epsilon_{in} = 10^{-7}, \ldots, 10^{-3}$. Here, we see, first, the same effect as in the previous analytical setup, namely that our approach leads especially to tighter intervals in noisy environments, i.e. for small N_2. Second, it seems that an unfortunate error

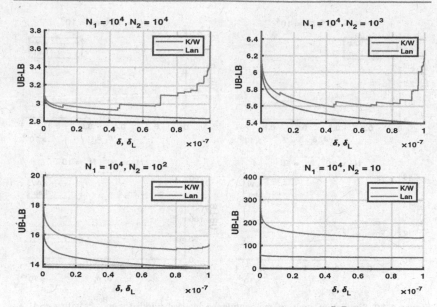

Figure 6.9 Simulation study of the confidence interval width for $q_\alpha^{\mathrm{BoF}_{t_1}}$ with $\epsilon = 0.1\%$, $\alpha = 0.5\%$ and varying δ, δ_L. For the approach of Lan et al. (2007b) we choose $\epsilon_{in} = 10^{-7}, \ldots, 10^{-3}$. This choice implies the corresponding parameters ϵ_{out} and δ_L. For comparison reasons we set $\delta = \delta_L$

splitting in Lan et al. (2007b) leads immediately to significant wider confidence intervals, cf. especially for $N_2 = 10^3, 10^4$. Indeed, this implies the effect that nearly the complete confidence level ϵ will be shifted to ϵ_{in} or ϵ_{out}. Then, no or a relatively small error budget remains for the contrary error source which seems to cause wider intervals.

Figure 6.10 indicates analogous effects for $\epsilon = 10\%$ and varying $\epsilon_{in} = 10^{-5}, \ldots, 10^{-1}$. Overall, it turns out that our approach performs under different CLT levels δ as well as different noise levels N_2 in this particular insurance setup quite well and outperforms the approach of Lan et al. (2007b).

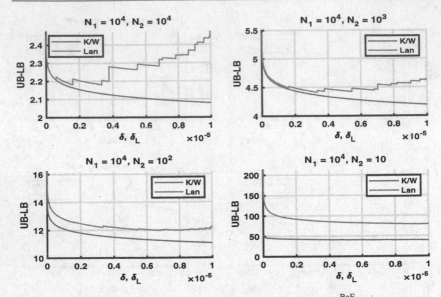

Figure 6.10 Simulation study of the confidence interval width for $q_\alpha^{\text{BoF}_{t1}}$ with $\epsilon = 10\%$, $\alpha = 0.5\%$ and varying δ, δ_L. For the approach of Lan et al. (2007b) we choose $\epsilon_{in} = 10^{-5}, \ldots, 10^{-1}$. This choice implies the corresponding parameters ϵ_{out} and δ_L. For comparison reasons we set $\delta = \delta_L$

Conclusion

Finally, we will close this thesis with some concluding remarks on our achievements and give a brief outlook for possible upcoming and so far unanswered and linked questions.

7.1 Achievements

In accordance to Bauer et al. (2010) resp. Bergmann (2011) we introduced, first, the general SCR problem from an insurance perspective. Second, we embedded this problem formulation into a probability theoretical one by introducing and focusing only on the necessary random variables, i.e. V, X and $X + \epsilon_{N_2}$. Based on this reformulation, we postulated two rather general problem formulations which were tackled in the course of this work, namely the moment-based and quantile-based case. Besides Rainforth et al. (2018) we were the first who investigated the moment-based case from such a general point of view, i.e. by addressing especially the vanishing bias problem in more detail and deriving optimal rates for a broad class of functions G. All other publications inspected – so far – mostly special examples of this class, like e.g. the probability of a large loss or lower partial moments, cf. Hong and Juneja (2009), Gordy and Juneja (2010), Broadie et al. (2011) and Liu et al. (2022).

In Chapter 3 we focused entirely on the moment-based problem formulation. Since the bias played a crucial role in obtaining optimal rates of convergence, we first investigated rates for this subproblem. Hence, we considered rates for a broad class of functions G. Then, based on the Fuk-Nagaev inequality and the complete convergence, we derived budget sequences which yield to our novel almost sure convergence results. Hence, we were able to show, on the one hand, the almost sure

M. Klein, *Nested Simulations: Theory and Application*, Mathematische Optimierung und Wirtschaftsmathematik | Mathematical Optimization and Economathematics, https://doi.org/10.1007/978-3-658-43853-1_7

convergence of moment-based estimators in general, cf. Theorem 3.4.1, and on the other hand, we provided also rates of convergence, cf. Theorem 3.4.4. Here, we proposed in the best case an a.s. rate of $N^{-1/3}$. For the mostly considered large loss problem we established, compared to the RMSE rate of Gordy and Juneja (2010) with $N^{-1/3}$, a remarkable optimal rate of $N^{-1/4}$ in an almost sure sense.

As a next step, in Chapter 4, we summarized our novel results on the quantile-based problem formulation. We also addressed, first, the vanishing quantile problem, i.e. $q_\alpha^{X+\epsilon_{N_2}} - q_\alpha^X$. Here, we derived two optimal rates of convergence: A rate of $-1/2$ under relatively weak assumptions and an order of -1 under strong smoothness assumptions on X in accordance to Gordy and Juneja (2010). In our following analysis we focused on the weak assumption set and derived the so far unanswered almost sure convergence and its corresponding rate of convergence based on the Hoeffding Theorem and the complete convergence. Already weak assumptions on the number of existing moments on V, i.e. $V \in L^3(\Omega, \mathcal{F}, \mathbb{P})$, and a non-flatness condition on F_X led to a remarkable rate of $N^{-1/4}$. This is compared to the optimal RMSE rate of $N^{-1/3}$ (cf. Gordy and Juneja (2010)) a notable result since $N^{-1/3}$ applies only under additional strong smoothness assumptions.

Further, we established in Chapter 5 a non parametric noise considering confidence interval approach for quantiles relying on a nested simulation. We figured out that existing non-noise considering methods, cf. David and Nagaraja (2004), led in noisy environments, i.e. small N_2, to a huge amount of misclassifications and thus they are not suitable for the problem at hand. Our novel achievement here is an asymptotic noise considering model-free confidence interval approach under relatively weak assumptions (i.e. $V \in L^3(\Omega, \mathcal{F}, \mathbb{P})$). A clear advantage of this approach, compared to Lan et al. (2007b), is a much easier parameterization. Furthermore, based on the Berry-Esseen Theorem, we were also able to quantify the underlying asymptotic error explicitly, a novelty so far.

In Chapter 6 we supported our developed theoretical results with an extensive numerical analysis. Thereby, we considered, first, various academic test examples and applied afterwards the confidence interval method to a simplified insurance model to underpin its usability to the SCR case at hand.

In total, we proved a so far and to the best of our knowledge missing puzzle piece in the theory of nested simulations, namely the almost sure convergence as well as their corresponding rates of convergence (cf. Table 1.1 and Table 1.2), on the one hand for the broad class of moment-based estimators and on the other hand for the financial/insurance relevant quantile estimators. Moreover, we developed an easy to parametrize noise considering confidence interval approach for quantiles.

7.2   **Outlook**

With regard to theoretical improvements, we consider the following points to be particularly noteworthy:

- Since the nested simulation problem formulation seems to be interpretable as a triangular array, i.e. an array of row-wise independent random variables, a new probability theoretical toolbox opens to address the described problems. In accordance with Hu et al. (1989) and Gut (1992) we see the potential to obtain also complete convergence results in the nested setting. Moreover, the triangular array theory seems to enable also immediately the chance to obtain with Billingsley (1995), Theorem 27.2, also a convergence in distribution. Based on that, it could be possible to derive uniform confidence intervals, an enhancement to our so far considered asymptotic intervals for fixed N_1.

- Aiming on further reduction of the previously weak assumptions for the a.s. results, especially the moment conditions on V, probability metrics and their properties seem to be helpful or at least worth mentioning. For example, the convergence of the distorted quantile $q_\alpha^{X+\epsilon_{N_2}}$ towards the quantile q_α^X can be characterized by the Lévy distance d_L, i.e.

$$\left| q_\alpha^{X+\epsilon_{N_2}} - q_\alpha^X \right| \leq K \cdot d_L(X + \epsilon_{N_2}, X),$$

for $K > 0$, if for each constant $\kappa, \tau > 0$

$$F_X\left(q_\alpha^X\right) - F_X\left(q_\alpha^X - \epsilon\right) \geq \kappa \cdot \epsilon, \qquad \forall \epsilon \in (0, \tau)$$

$$F_X\left(q_\alpha^X + \epsilon\right) - F_X(q_\alpha^X) \geq \kappa \cdot \epsilon, \qquad \forall \epsilon \in (0, \tau)$$

holds, instead of 'strong' moment assumptions on V. Maybe this approach could also be further generalized for moment-based considerations with other or inclined probability metrics.

Regarding to numerical resp. practical improvements, we consider the upcoming points to be particularly noteworthy:

- Our considered insurance model in the numerical investigations was a rather simple one with only a few risk factors and without default events. Hence, for insurers it would be certainly of particular interest to conduct further confidence interval analyses based on a more realistic ALM model, allowing for example default events during the policy duration. Here, from an academic perspective the models of Gerstner (2008) resp. Krah et al. (2018) seem to be helpful to obtain a more realistic framework.

- In an environment where analytical pricing formulas are impossible to obtain, like the present SCR case at hand, it is necessary to resort brute force methods like nested simulations. It should be highlighted that these are the only obtainable reference values which allow a noise control, the main problem at hand. Hence, our almost sure convergence and especially the corresponding budget sequence splittings set the minimum requirements for the composition of datasets in machine learning approaches for risk management tasks. If the addressed requirements are fulfilled, it is certainly interesting to learn either the BoF_{t_1} or DTV directly in order to bypass the computationally burdensome nested simulation. Hence, it is interesting if such non-linear methods further improve the linear LSMC resp. replication approaches. First, noteworthy publications are e.g. Hejazi and Jackson (2017), Krah et al. (2020a,b).

A. Calculations for the Academic Test Example 6.1.2

A

In Example 6.1.2 we consider the following simulation framework:

$$Z \sim \mathcal{N}(0, 1),$$
$$V|Z \sim \mathcal{LN}(0, Z^2/t)$$

Indeed, the fact that $V|Z$ is lognormal distributed leads obviously to

$$X = \mathbb{E}[V|Z] = \mathbb{E}\left[\exp\left(\frac{|Z| \cdot Y}{\sqrt{t}}\right)\Big|Z\right] = e^{Z^2 \cdot \frac{1}{2t}},$$

with Y standard normal distributed and independent from Z.

A.1   Moment-Based Case $G(X) = X^2$

Now, the moment-based case with $G(X) = X^2$ leads directly to

$$G(X) = G\left(\mathbb{E}[V|Z]\right) = \left(\mathbb{E}[V|Z]\right)^2 = e^{Z^2 \cdot \frac{1}{t}}.$$

In order to calculate an analytical solution of (1.10), it should be emphasized, first, that Z^2 is χ_1^2-distributed and that, second, $\mathbb{E}\left[e^{Z^2 \cdot \frac{1}{t}}\right]$ describes the classical form of a moment generating function. The moment generating function for Z^2 is then generally given by

$$M_{Z^2}(c) := \mathbb{E}\left[e^{c \cdot Z^2}\right] = \frac{1}{\sqrt{1 - 2 \cdot c}},$$

© The Editor(s) (if applicable) and The Author(s), under exclusive license to Springer Fachmedien Wiesbaden GmbH, part of Springer Nature 2024
M. Klein, *Nested Simulations: Theory and Application*, Mathematische Optimierung und Wirtschaftsmathematik | Mathematical Optimization and Economathematics, https://doi.org/10.1007/978-3-658-43853-1

for $c < 1/2$. If $t > 2$ holds, then, based on the moment generating function for χ_1^2-distributed random variables, we obtain a solution for γ, i.e.

$$\gamma = \mathbb{E}[G(X)] = \mathbb{E}\left[G\left(\mathbb{E}[V|Z]\right)\right] = \mathbb{E}\left[e^{Z^2 \cdot \frac{1}{t}}\right]$$
$$= \frac{1}{\sqrt{1 - 2 \cdot \frac{1}{t}}}.$$

For $(A2,p)$ we immediately obtain

$$||G(X)||_{L^p} = \mathbb{E}\left[|G(X)|^p\right]^{1/p} = \mathbb{E}\left[X^{2p}\right]^{1/p}$$
$$= \mathbb{E}\left[\mathbb{E}[V|Z]^{2p}\right]^{1/p} = \mathbb{E}\left[\left(e^{Z^2 \cdot \frac{1}{2t}}\right)^{2p}\right]^{1/p}$$
$$= \mathbb{E}\left[e^{Z^2 \cdot \frac{p}{t}}\right]^{1/p}.$$

Now, according to the moment generating function of Z^2, it should be highlighted that $(A2,p)$ applies if $t > 2p$ holds, i.e.

$$\mathbb{E}\left[e^{Z^2 \cdot \frac{p}{t}}\right]^{1/p} = \left(\frac{1}{\sqrt{1 - 2 \cdot \frac{p}{t}}}\right)^{1/p} < \infty,$$

otherwise $||G(X)||_{L^p} = \infty$ and thus also $(A1,p)$ does not hold according to Proposition 3.1.1 (3).

A.2   Quantile-Based Case

In a next step we obtain the Value-at-Risk for $\alpha \in (0, 1)$ by considering, first of all, the before calculated conditional expectation

$$VaR_\alpha \left(\mathbb{E}[V|Z]\right) = -q_\alpha \left(\mathbb{E}[V|Z]\right) = -q_\alpha \left(e^{Z^2 \cdot \frac{1}{2t}}\right)$$

and using, second of all, the fact that Z is standard normal distributed which yields

$$-q_\alpha \left(e^{Z^2 \cdot \frac{1}{2t}} \right) = -\exp \left(\frac{\Phi^{-1} \left(\frac{1}{2} + \frac{\alpha}{2} \right)^2}{2t} \right),$$

whereby Φ^{-1} denotes the inverse of the standard normal cdf.

In order to obtain optimal rates in the quantile case, we have to verify Assumption $(B1,p)$ and thus when $||V||_{L^p} < \infty$ applies. In general, it holds

$$||V||_{L^p}^p = \mathbb{E}\left[V^p \right] = \mathbb{E}\left[\mathbb{E}\left[V^p | Z \right] \right].$$

Since $V|Z \sim \mathcal{LN}(0, Z^2/t)$, we immediately obtain the p-th moment, i.e.

$$\mathbb{E}\left[\mathbb{E}\left[V^p | Z \right] \right] = \mathbb{E}\left[\exp \left(\frac{p^2 \cdot Z^2}{2t} \right) \right] = \mathbb{E}\left[\exp \left(\frac{p^2}{2t} \cdot Z^2 \right) \right].$$

Once again, the fact that Z^2 is χ_1^2-distributed leads in combination with the corresponding moment generating function to

$$\mathbb{E}\left[\exp \left(\frac{p^2}{2 \cdot t} \cdot Z^2 \right) \right] = \frac{1}{\sqrt{1 - 2 \cdot \frac{p^2}{2t}}} = \frac{1}{\sqrt{1 - \frac{p^2}{t}}}.$$

Hence, it follows directly that $(B1,p)$ and thus $||V||_{L^p} < \infty$ holds, if $t > p^2$ applies.

B. Derivative of the Binomial Distribution and Consequences

B

Lemma B.1 *(Partial derivative of the binomial distribution regarding to p)*
The first partial derivative of the binomial distribution with respect to the underlying success probability p is given by

$$\frac{\partial}{\partial p} \mathcal{B}_{N,p}(k) = -N \binom{N-1}{k} p^k (1-p)^{N-1-k}, \tag{B.1}$$

for $N, k \in \mathbb{N}_0$, $k \leq N$ and $p \in [0, 1]$.

Proof.
Let $X \sim \mathcal{B}_{N,p}$ and let $\mathbb{P}(X = k) = \binom{N}{k} p^k (1-p)^{N-k}$ define the probability mass function (pmf) of the underlying random variable X. The logarithm of the pmf can be rewritten into

$$\log\left(\mathbb{P}(X = k)\right) = \log\left(\binom{N}{k} p^k (1-p)^{N-k}\right)$$
$$= \log\left(\binom{N}{k}\right) + k \cdot \log(p) + (N - k) \cdot \log(1 - p)$$

and hence the partial derivative regarding to p is simply

$$\frac{\partial}{\partial p} \log\left(\mathbb{P}(X = k)\right) = \frac{k}{p} - \frac{N - k}{1 - p}.$$

© The Editor(s) (if applicable) and The Author(s), under exclusive license to Springer Fachmedien Wiesbaden GmbH, part of Springer Nature 2024
M. Klein, *Nested Simulations: Theory and Application*, Mathematische Optimierung und Wirtschaftsmathematik | Mathematical Optimization and Economathematics, https://doi.org/10.1007/978-3-658-43853-1

The chain rule from analysis leads to

$$\frac{\partial}{\partial p} \mathbb{P}(X = k) = \mathbb{P}(X = k) \frac{\partial}{\partial p} \log(\mathbb{P}(X = k))$$

$$= \mathbb{P}(X = k) \cdot \left(\frac{k}{p} - \frac{N - k}{1 - p}\right)$$

$$= \binom{N}{k} p^k (1 - p)^{N-k} \left(\frac{k}{p} - \frac{N - k}{1 - p}\right). \tag{B.2}$$

Having considered the partial derivative of the pmf, this should be extended to the partial derivative of the cdf in a next step. The continuity of the binomial cdf and (B.2) thus lead to

$$\frac{\partial}{\partial p} \mathcal{B}_{N,p}(k) = \frac{\partial}{\partial p} \sum_{i=0}^{k} \binom{N}{i} p^i (1 - p)^{N-i} = \sum_{i=0}^{k} \underbrace{\frac{\partial}{\partial p} \binom{N}{i} p^i (1 - p)^{N-i}}_{\mathbb{P}(X=i)}$$

$$= \sum_{i=0}^{k} \binom{N}{i} p^i (1 - p)^{N-i} \cdot \left(\frac{i}{p} - \frac{N - i}{1 - p}\right)$$

$$= \sum_{i=0}^{k} \binom{N}{i} i \cdot p^{i-1} (1 - p)^{N-i} - \binom{N}{i} (N - i) p^i (1 - p)^{N-1-i}$$

$$= \sum_{i=0}^{k} \binom{N}{i} i \cdot p^{i-1} (1 - p)^{N-i} - (i + 1) \binom{N}{i+1} p^{(i+1)-1} (1 - p)^{N-(i+1)}$$

$$= -(k + 1) \binom{N}{k+1} p^{(k+1)-1} (1 - p)^{N-(k+1)} = -\binom{N}{k} (N - k) p^k (1 - p)^{N-1-k}$$

$$= -N \binom{N-1}{k} p^k (1 - p)^{N-1-k} = -N \cdot \mathbb{P}(\widehat{X} = k),$$

for $\widehat{X} \sim \mathcal{B}_{N-1,p}$. From the fourth to the fifth line we used the fact that the sum at hand is telescopic.   □

Remark B.2

Indeed, a direct consequence of Lemma B.1 is that the binomial distribution function is monotonically decreasing in p because $\frac{\partial}{\partial p} \mathcal{B}_{N,p}(k) \leq 0$ holds for $N, k \in \mathbb{N}_0$, $k \leq N$ and $p \in [0, 1]$.

References

Andradóttir, S. and Glynn, P. (2016). Computing Bayesian Means Using Simulation. *ACM Transactions on Modeling and Computer Simulation*, 26(2):1–26.

Andreatta, G. and Corradin, S. (2003). Valuing the Surrender Options Embedded in a Portfolio of Italian Life Guaranteed Participating Policies: A Least Squares Monte Carlo Approach. *Working Paper*.

Aven, T. (2016). Risk Assessment and Risk Management: Review of Recent Advances on their Foundation. *European Journal of Operational Research*, 253(1):1–13.

Bacinello, A. R., Biffis, E., and Millossovich, P. (2009). Pricing Life Insurance Contracts with Early Exercise Features. *Journal of Computational and Applied Mathematics*, 233(1):27–35.

Bacinello, A. R., Biffis, E., and Millossovich, P. (2010). Regression-Based Algorithms for Life Insurance Contracts with Surrender Guarantees. *Quantitative Finance*, 10(9):1077–1090.

Baione, F., De Angelis, P., and Fortunati, A. (2006). On a Fair Value Model for Participating Life Insurance Policies. *Investment Management and Financial Innovations*, 3(2):105–115.

Bauer, D., Bergmann, D., and Kiesel, R. (2012). On the Risk-Neutral Valuation of Life Insurance Contracts with Numerical Methods in View. *Astin Bulletin*, 42(2):65–95.

Bauer, D. and Ha, H. (2022). A Least-Squares Monte Carlo Approach to the Estimation of Enterprise Risk. *Finance and Stochastics*, 26(3):417–459.

Bauer, D., Reuss, A., and Singer, D. (2010). On the Calculation of the Solvency Capital Requirement Based on Nested Simulations. *Astin Bulletin*, 40(1):453–499.

Becker, T., Cottin, C., Fahrenwaldt, M., Hamm, A.-M., Nörtemann, S., and Weber, S. (2014). Market Consistent Embedded Value – Eine praxisorientierte Einführung. *Der Aktuar*, 1. URL: https://www.insurance.uni-hannover.de/fileadmin/house-of-insurance/Publications/2014/Market_Consistent_Embedded_Value.pdf, accessed January 2021.

Bentkus, V. and Götze, F. (1996). The Berry-Esseen Bound for Student's Statistic. *The Annals of Probability*, 24(1):491–503.

Bergmann, D. (2011). *Nested Simulations in Life Insurance*. PhD thesis, Universität Ulm.

Bernard, A. E. and Lemieux, C. (2008). Fast Simulation of Equity-Linked Life Insurance Contracts with a Surrender Option. *Proceedings of the 2008 Winter Simulation Conference*, pages 444–452.

Berry, A. (1941). The Accuracy of the Gaussian Approximation to the Sum of Independent Variates. *Transactions of the American Mathematical Society*, 49:122–136.

M. Klein, *Nested Simulations: Theory and Application*, Mathematische Optimierung und Wirtschaftsmathematik | Mathematical Optimization and Economathematics, https://doi.org/10.1007/978-3-658-43853-1

Berti, P., Pratelli, L., and Rigo, P. (2004). Limit Theorems for a Class of Identically Distributed Random Variables. *Annals of Probability*, 32(3):2029–2052.

Beutner, E., Schweizer, J., and Pelsser, A. (2014). Fast Convergence of Regress-Later Estimates in Least Squares Monte Carlo. arXiv:1309.5274, pages 1–25.

Billingsley, P. (1995). *Probability and Measure*. Wiley Series in Probability and Mathematical Statistics. Wiley, New York, 3rd edition.

Boos, D. (1985). A Converse on Scheffés Theorem. *The Annals of Statistic*, 13(1):423–427.

Broadie, M. and Cao, M. (2008). Improved Lower and Upper Bound Algorithms for Pricing American Options by Simulation. *Quantitative Finance*, 8(8):845–861.

Broadie, M., Du, Y., and Moallemi, C. C. (2011). Efficient Risk Estimation via Nested Sequential Simulation. *Management Science*, 57(6):1172–1194.

Broadie, M., Du, Y., and Moallemi, C. C. (2015). Risk Estimation via Regression. *Operations Research*, 63(5):979–1244.

Carriere, J. F. (1996). Valuation of the Early-Exercise Price for Options Using Simulations and Nonparametric Regression. *Insurance: Mathematics and Economics*, 19(1):19–30.

CFO Forum (2009). Market Consistent Embedded Value Principles. Technical Report, European Insurance CFO Forum. URL: http://www.cfoforum.eu/downloads/MCEV_Principles_and_Guidance_October_2009.pdf, accessed January 2021.

CFO Forum (2016). Market Consistent Embedded Value Principles. Technical Report, European Insurance CFO Forum. URL: http://www.cfoforum.eu/downloads/CFO-Forum_MCEV_Principles_and_Guidance_April_2016.pdf, accessed January 2021.

Chen, W. and Skoglund, J. (2012). Cashflow Replication with Mismatch Constraints. *Journal of Risk*, 14(4):115–128.

Clément, E., Lamberton, D., and Protter, P. (2002). An Analysis of a Least Squares Regression Method for American Option Pricing. *Finance and Stochastics*, 6:449–471.

Daul, S. and Vidal, E. G. (2009). Replication of Insurance Liabilities. *RiskMetrics Journal*, 9(1):79–96.

David, H. and Nagaraja, H. (2004). *Order Statistics*. Wiley Series in Probability and Statistics. Wiley, New Jersey, 3rd edition.

Egloff, D., Kohler, M., and Todorovic, N. (2007). A Dynamic Look-Ahead Monte Carlo Algorithm for Pricing Bermudan Options. *The Annals of Applied Probability*, 17(4):1138–1171.

Elstrodt, J. (2018). *Maß- und Integrationstheorie*, volume 8. Springer Spektrum, Berlin.

Esseen, C. (1942). On the Liapounoff Limit of Error in the Theory of Probability. *Ark.Mat. Astr. Fys.*, 28 A:1–19.

Esseen, C. (1956). A Moment Inequality with an Application to the Central Limit Theorem. *Scandinavian Actuarial Journal*, 1956(2):160–170.

European Parliament and European Council (2009). Directive 2009/138/EC on the Takingup and Pursuit of the Business of Insurance and Reinsurance (Solvency II).

Feng, R. and Peng, L. (2021). Sample Recycling Method—A New Approach to Efficient Nested Monte Carlo Simulations. *Working Paper*. URL: https://arxiv.org/pdf/2106.06028.pdf, accessed January 2022.

Föllmer, H. and Schied, A. (2016). *Stochastic Finance: An Introduction in Discrete Time*. De Gruyter, Berlin, 4th edition.

Gerstner, T., Griebel, M., Holtz, M., Goschnick, R., and Haep, M. (2008). A General Asset-Liability Management Model for the Efficient Simulation of Portfolios of Life Insurance Policies. *Insurance: Mathematics and Economics*, 42(2):704–716.

Glasserman, P. and Yu, B. (2004). Simulation for American Options: Regression Now or Regression Later. *Monte Carlo and Quasi-Monte Carlo Methods 2002*, pages 213–226.

Glynn, P. and Lee, S.-H. (2003). Computing the Distribution Function of a Conditional Expectation via Monte Carlo: Discrete Conditioning Spaces. *ACM Transactions on Modeling and Computer Simulation*, 13(3):238–258.

Gordy, M. and Juneja, S. (2010). Nested Simulation in Portfolio Risk Management. *Management Science*, 56(10):1833–1848.

Gut, A. (1985). On Complete Convergence in the Law of Large Numbers for Subsequences. *The Annals of Probability*, 13(4):1286–1291.

Gut, A. (1992). Complete Convergence for Arrays. *Periodica Mathematica Hungarica*, 25(1):51–75.

Hancock, J., Huber, P., and Koch, P. (2001). *The Economics of Insurance: How Insurers Create Value for Shareholders*. Swiss Reinsurance Company.

Hejazi, S. and Jackson, K. (2017). Efficient Valuation of SCR via a Neural Network Approach. *Journal of Computational and Applied Mathematics*, 313(2017):427–439.

Hieber, P., Natolski, J., and Werner, R. (2019). Fair Valuation of Cliquet-Style Return Guarantees in (Homogeneous and) Heterogeneous Life Insurance Portfolios. *Scandinavian Actuarial Journal*, 2019(6):478–507.

Hoeffding, W. (1963). Probability Inequalities for Sums of Bounded Random Variables. *Journal of the American Statistical Association*, 58(301):13–30.

Holtz, M. (2008). *Sparse Grid Quadrature in High Dimensions with Applications in Finance and Insurance*. PhD thesis, Rheinischen Friedrich-Wilhelms-Universität Bonn.

Hong, J., Juneja, S., and Liu, G. (2017). Kernel Smoothing for Nested Estimation with Application to Portfolio Risk Measurement. *Operations Research*, 65(3):657–673.

Hong, L. and Juneja, S. (2009). Estimating the Mean of a Non-Linear Function of Conditional Expectation. *Proceedings of the 2009 Winter Simulation Conference (WSC)*, pages 1223–1236.

Hsu, P. and Robins, H. (1947). Complete Convergence and the Law of Large Numbers. *Proceedings of the National Academy of Sciences*, 33(2):25–31.

Hu, T.-C., Móricz, F., and Taylor, R. L. (1989). Strong Laws of Large Numbers for Arrays of Rowwise Independent Random Variables. *Acta Mathematica Hungarica*, 54(1–2):153–162.

Hult, H., Lindskog, F., Hammarlid, O., and Rehn, C. J. (2012). *Risk and Portfolio Analysis*. Springer Series in Operations Research and Financial Engineering. Springer, New York.

Investment Committee of DAV (2015). Zwischenbericht zur Kalibrierung und Validierung spezieller ESG unter Solvency II. Technical Report, Investment Committee of Deutsche Aktuarvereinigung.

Kalberer, T. (2012a). Stochastic Determination of the Value at Risk for a Portfolio of Assets and Liabilities – Typical and New Approaches and Implied Estimation Errors (Part 1). *Der Aktuar*, pages 12–22.

Kalberer, T. (2012b). Stochastic Determination of the Value at Risk for a Portfolio of Assets and Liabilities – Typical and New Approaches and Implied Estimation Errors (Part 2). *Der Aktuar*.

Kay, J. and King, M. (2020). *Radical Uncertainty: Decision-Making for an Unknowable Future*. The Bridge Street Press, London.

Klein, M. and Werner, R. (2023a). Asymptotic Non Parametric Confidence Intervals for Quantiles based on Nested Simulations. *Working Paper*.

Klein, M. and Werner, R. (2023b). On Almost Sure Convergence of Moment-Based Nested Simulation Estimators. *Working Paper*.

Klein, M. and Werner, R. (2023c). On Almost Sure Convergence of Quantile-Based Nested Simulation Estimators. *Working Paper*.

Klenke, A. (2014). *Probability Theory*. Universitext. Springer, London, 2nd edition.

Krah, A. S. (2020). *Least-Squares Monte Carlo Methods in the Life Insurance Sector*. PhD thesis, Technische Universität Kaiserslautern.

Krah, A. S., Nikolić, Z., and Korn, R. (2018). A Least-Squares Monte Carlo Framework in Proxy Modeling of Life Insurance Companies. *Risks*, 6(62):1–26.

Krah, A. S., Nikolić, Z., and Korn, R. (2020a). Least-Squares Monte Carlo for Proxy Modeling in Life Insurance: Neural Networks. *Risks*, 8(4):1–21.

Krah, A. S., Nikolić, Z., and Korn, R. (2020b). Machine Learning in Least-Squares Monte Carlo Proxy Modeling of Life Insurance Companies. *Risks*, 8(1):1–79.

Lan, H., Nelson, B., and Staum, J. (2007a). A Confidence Interval for Tail Conditional Expectation via Two-Level Simulation. *Proceedings of the 2007 Winter Simulation Conference*, pages 949–957.

Lan, H., Nelson, B., and Staum, J. (2007b). Two-Level Simulations for Risk Management. *Proceedings of the 2007 INFORMS Simulation Society Research Workshop*, pages 102–107.

Lan, H., Nelson, B., and Staum, J. (2010). A Confidence Interval Procedure for ExpectedShortfall Risk Measurement via Two-Level Simulation. *Operations Research*, 58(5):1481–1490.

Lee, S.-H. (1998). *Monte Carlo Computation of Conditional Expectation Quantiles*. PhD thesis, Stanford University.

Liu, G., Wang, S., and Zhang, K. (2022). Bootstrap-based Budget Allocation for Nested Simulation. *Operations Research*, 70(2):1128–1142.

Longstaff, F. A. and Schwartz, E. S. (2001). Valuing American Options by Simulation: A Simple Leas-Squares Approach. *The Review of Financial Studies*, 14(1):113–147.

Majerak, D., Nowak, W., and Zieba, W. (2005). Conditional Strong Law of Large Number. *International Journal of Pure and Applied Mathematics*, 20(2):143–156.

McNeil, A., Embrechts, P., and Frey, R. (2015). *Quantitative Risk Management: Concepts, Techniques and Tools*. Princeton Series in Finance. Princeton University Press, New Jersey, Revised edition.

Nagaev, S. V. (1979). Large Deviations of Sums of Independent Random Variables. *The Annals of Probability*, 7(5):745–789.

Natolski, J. (2018). *Replizierende Portfolios in der Lebensversicherung*. Mathematische Optimierung und Wirtschaftsmathematik | Mathematical Optimization and Economathematics. Springer Spektrum, Wiesbaden.

Natolski, J. and Werner, R. (2014). Mathematical Analysis of Different Approaches for Replicating Portfolios. *European Actuarial Journal*, 2:411–435.

Natolski, J. and Werner, R. (2015). Improving Optimal Terminal Value Replicating Portfolios. In Glau, K., Scherer, M., and Zagst, R., editors, *Innovations in Quantitative Risk Management*, pages 289–301. Springer International Publishing.

Novinger, W. P. (1972). Mean Convergence in L^p Spaces. *Proceedings of the American Mathematical Society*, 34(2):627–628.

Oechslin, J., Aubry, O., Aellig, M., Kappeli, A., Bronnimann, D., Tandonnet, A., and Valois, G. (2007). Replicating Embedded Options. *Life & Pensions Risk*, pages 47–52.

Oezkan, F., Seemann, A., and Wendel, S. (2011). Replizierende Portfolios in der Lebensversicherung. In Bennemann, C., Oehlenberg, L., and Stahl, G., editors, *Handbuch Solvency II : von der Standardformel zum internen Modell, vom Governance-System zu den MaRisk VA*, pages 163–176, Stuttgart. Schäffer-Poeschel.

Pelsser, A. (2003). Pricing and Hedging Guaranteed Annuity Options via Static Option Replication. *Insurance: Mathematics and Economics*, 33(2):283–296.

Pelsser, A. and Schweizer, J. (2016). The Difference Between LSMC and Replicating Portfolio in Insurance Liability Modeling. *European Actuarial Journal*, 6:441–494.

Petrov, V. V. (1975). *Sums of Independent Random Variables*. Ergebnisse der Mathematik und ihre Grenzgebiete. Springer, Berlin.

Rainforth, T., Cornish, R., Yang, H., Warrington, A., and Wood, F. (2018). On nesting Monte Carlo estimators. In Dy, J. and Krause, A., editors, *Proceedings of the 35th International Conference on Machine Learning*, volume 80 of *Proceedings of Machine Learning Research*, pages 4267–4276. PMLR.

Rainforth, T., Cornish, R., Yang, H., and Wood, F. (2016). On the Pitfalls of Nested Monte Carlo. *NeurIPS Workshop on Advances in Approximate Bayesian Inference*.

Resnick, S. (2005). *A Probability Path*, volume 5. Birkhäuser, Boston.

Schweizer, J. (2016). *Portfolio Replication and Least Squares Monte Carlo with Application to Insurance Risk Management*. PhD thesis, Maastricht University.

Seemann, A. (2009). *Replizierende Portfolios in der deutschen Lebensversicherung*. PhD thesis, University of Ulm.

Serfling, R. J. (1980). *Approximation Theorems of Mathematical Statistics*. Wiley Series in Probability and Statistics. John Wiley & Sons, Inc., New York.

Sweeting, T. (1986). On a Converse to Scheffés Theorem. *The Annals of Statistic*, 14(3):1252–1256.

Tsitsiklis, J. N. and van Roy, B. (2001). Regression Methods for Pricing Complex American-Style Options. *IEEE Transactions on Neural Networks*, 12(4):694–703.

Yuan, D., An, J., and Wu, X. (2010). Conditional Limit Theorems for Conditionally Negatively Associated Random Variables. *Monatshefte für Mathematik*, 161:449–473.

Yuan, D., Wei, L., and Lei, L. (2014). Conditional Central Limit Theorems for a Sequence of Conditional Independent Random Variables. *Journal of the Korean Mathematical Society*, 51(1):1–15.

Zanger, D. (2018). Convergence of a Least-Squares Monte Carlo Algorithm for American Option Pricing with Dependent Sample Data. *Mathematical Finance*, 28(1):447–479.

Printed in the United States
by Baker & Taylor Publisher Services